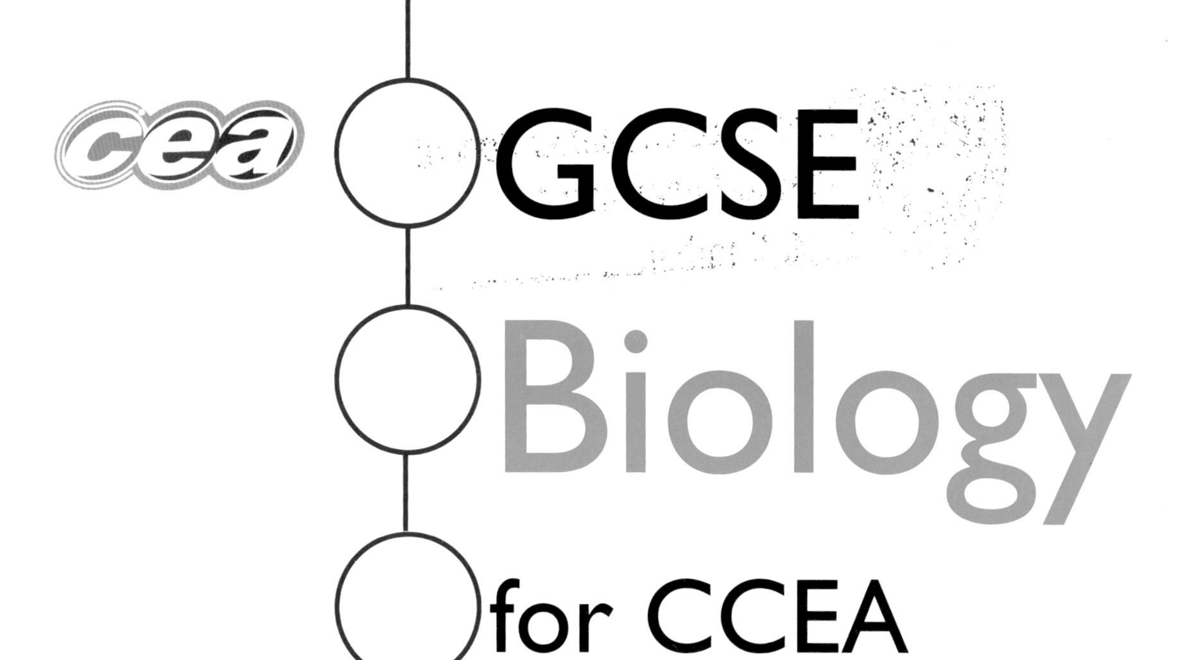

second edition

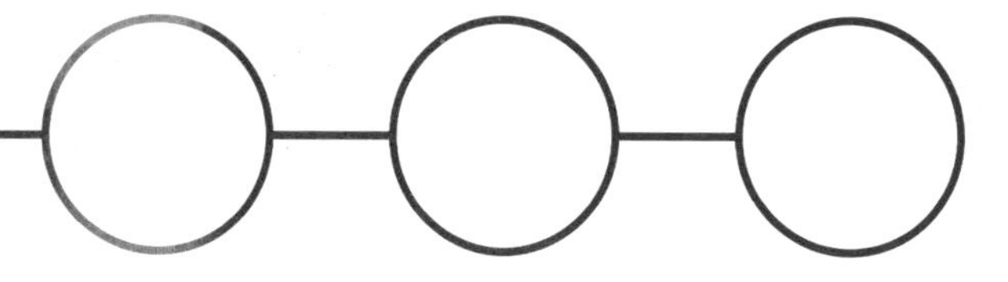

James Napier

HODDER EDUCATION

Endorsed by CCEA on 18th July 2011 If in any doubt about the continuing currency of CCEA endorsement, please contact Heather Clarke at CCEA, 29 Clarendon Road, Belfast BT1 3BG.

Acknowledgements
The Publisher would like to thank the following for permission to reproduce copyright material:

Photo credits
p.22 volff – Fotolia; **p.23** *t* Rosenfeld/Photolibrary Group; **p.23** *b* Duöan Zidar – Fotolia; **p.24** ANDREW MCCLENAGHAN/SCIENCE PHOTO LIBRARY; **p.33** © N. McKnight; **p.36** STEVE GSCHMEISSNER/SCIENCE PHOTO LIBRARY; **p.42** *l* BSIP, CHASSENET/SCIENCE PHOTO LIBRARY, *r* BSIP, CHASSENET/SCIENCE PHOTO LIBRARY; **p.45** SCIENCE PHOTO LIBRARY; **p.46** SUE FORD/SCIENCE PHOTO LIBRARY; **p.47** *l* © Nigel Cattlin/Alamy, *r* © Nigel Cattlin/Alamy; **p.48** Elenathewise – Fotolia; **p.52** © Realimage/Alamy; **p.53** Jason Smalley/Nature PL/Specialist Stock; **p.68** © Wildscape/Alamy; **p.69** © Paul Lindsay/Alamy; **p.84** *t* A. BARRINGTON BROWN/SCIENCE PHOTO LIBRARY, *b* PATRICK LANDMANN/SCIENCE PHOTO LIBRARY; **p.89** GIRAND/SCIENCE PHOTO LIBRARY; **p.91** BSIP, RAGUET/SCIENCE PHOTO LIBRARY; **p.93** Hulton Archive/Getty Images; **p.107** EDELMANN/SCIENCE PHOTO LIBRARY; **p.108** *l* TEK IMAGE/SCIENCE PHOTO LIBRARY, *r* ADAM HART-DAVIS/SCIENCE PHOTO LIBRARY; **p.110** AJ PHOTO/SCIENCE PHOTO LIBRARY; **p.114** *l* L.WILLATT, EAST ANGLIAN REGIONAL GENETICS SERVICE/SCIENCE PHOTO LIBRARY, *r* CNRI/SCIENCE PHOTO LIBRARY; **p.121** © PhotoStockFile/Alamy; **p.122** Bob Thomas/Popperfoto/Getty Images; **p.123** akg-images; **p.129** *l* SCIENCE PHOTO LIBRARY, *r* SCIENCE PHOTO LIBRARY; **p.130** Gina Sanders – Fotolia; **p.143** JOHN DURHAM/SCIENCE PHOTO LIBRARY; **p.145** *t* Oxford Science Archive/HIP/TopFoto, *b* NATIONAL LIBRARY OF MEDICINE/SCIENCE PHOTO LIBRARY; **p.147** © Kuttig – People/Alamy; **p.148** TEK IMAGE/SCIENCE PHOTO LIBRARY

t = top, *b* = bottom, *l* = left, *r* = right, *m* = middle

Every effort has been made to trace all copyright holders, but if any have been inadvertantly overlooked, the Publisher will be pleased to make the necessary arrangements at the first opportunity.

Although every effort has been made to ensure that website addresses are correct at time of going to press, Hodder Education cannot be held responsible for the content of any website mentioned. It is sometimes possible to find a relocated web page by typing in the address of the home page for a website in the URL window of your browser.

Hachette UK's policy is to use papers that are natural, renewable and recyclable products and made from wood grown in sustainable forests. The logging and manufacturing processes are expected to conform to the environmental regulations of the country of origin.

Orders: please contact Bookpoint Ltd, 130 Milton Park, Abingdon, Oxon OX14 4SB. Telephone: (44) 01235 827720. Fax: (44) 01235 400454. Lines are open 9.00–17.00, Monday to Saturday, with a 24-hour message answering service. Visit our website at www.hoddereducation.co.uk

First published in 2011 by
Hodder Education
An Hachette UK Company,
338 Euston Road
London NW1 3BH

Impression number	5	4	3	
Year	2015	2014	2013	2012

Cover photo EMMELINE WATKINS/SCIENCE PHOTO LIBRARY

Illustrations by Barking Dog Art
Typeset in 12/14 Bembo by Tech-Set Ltd., Gateshead, Tyne and Wear.
Printed in Dubai.

A catalogue record for this title is available from the British Library

ISBN 978 0340 983805

Contents

Preface iv

Unit 1

1 Cells 1
2 Photosynthesis and Plants 7
3 Nutrition and Health 18
4 Digestion and Enzymes 26
5 Breathing and the Respiratory System 33
6 The Nervous System and Hormones 39
7 Ecological Relationships and Energy Flow 51

Unit 2

8 Osmosis and Plant Transport 74
9 Chromosomes, Genes and DNA 81
10 Cell Division and Genetics 87
11 Reproduction, Fertility and Contraception 106
12 Applied Genetics 113
13 Variation and Selection 120
14 The Circulatory System 126
15 Microorganisms, Defence Against Disease, Medicine and Drugs 135
16 Controlled Assessment 153

Index 157

Preface

The GCSE Science for CCEA series comprises three books: GCSE Biology for CCEA, GCSE Chemistry for CCEA and GCSE Physics for CCEA, which together cover all aspects of the material needed for students following the CCEA GCSE specifications in:

* Science: Double Award
* Science: Biology
* Science: Chemistry
* Science: Physics

GCSE Biology for CCEA covers all the material relating to the biology component of the CCEA Science Double Award, together with the additional material required for the CCEA Science: Biology specification.

Dr James Napier is a teacher, senior examiner and author.

Identifying Specification and Tier

The material required for each specification and tier is clearly identified using the following colour code:

All material not on a tinted background is required for foundation tier students following either the GCSE Double Award Science or the GCSE Biology specifications. All foundation tier material can also be assessed at higher tier.

Material required for the higher tier students following either the GCSE Double Award Science or the GCSE Biology specification is identified with a green tinted background.

Material required for foundation tier students following the GCSE Biology specification is identified with a blue tinted background.

Material required for higher tier students following the GCSE Biology specification is identified with a red tinted background.

Material with a yellow tint on page 57 is for Double Award Science not required for GCSE Biology.

Controlled Assessment of Practical Skills

During your course you will be required to carry out a number of controlled assessment tasks.

Double Award Science students complete **two** controlled assessment tasks from a choice of six supplied by CCEA at the start of the GCSE course. The two tasks must come from different subject areas within the specification. So, for example, they cannot both come from the biology section of the Double Award Science specification.

For GCSE Biology students, CCEA sets two comparable tasks at the start of the GCSE course. Candidates must take at least **one** of these controlled assessment tasks in the course of the two years.

1 Cells

Living organisms are made up of microscopic units called cells. Figure 1.1 shows typical animal and plant cells.

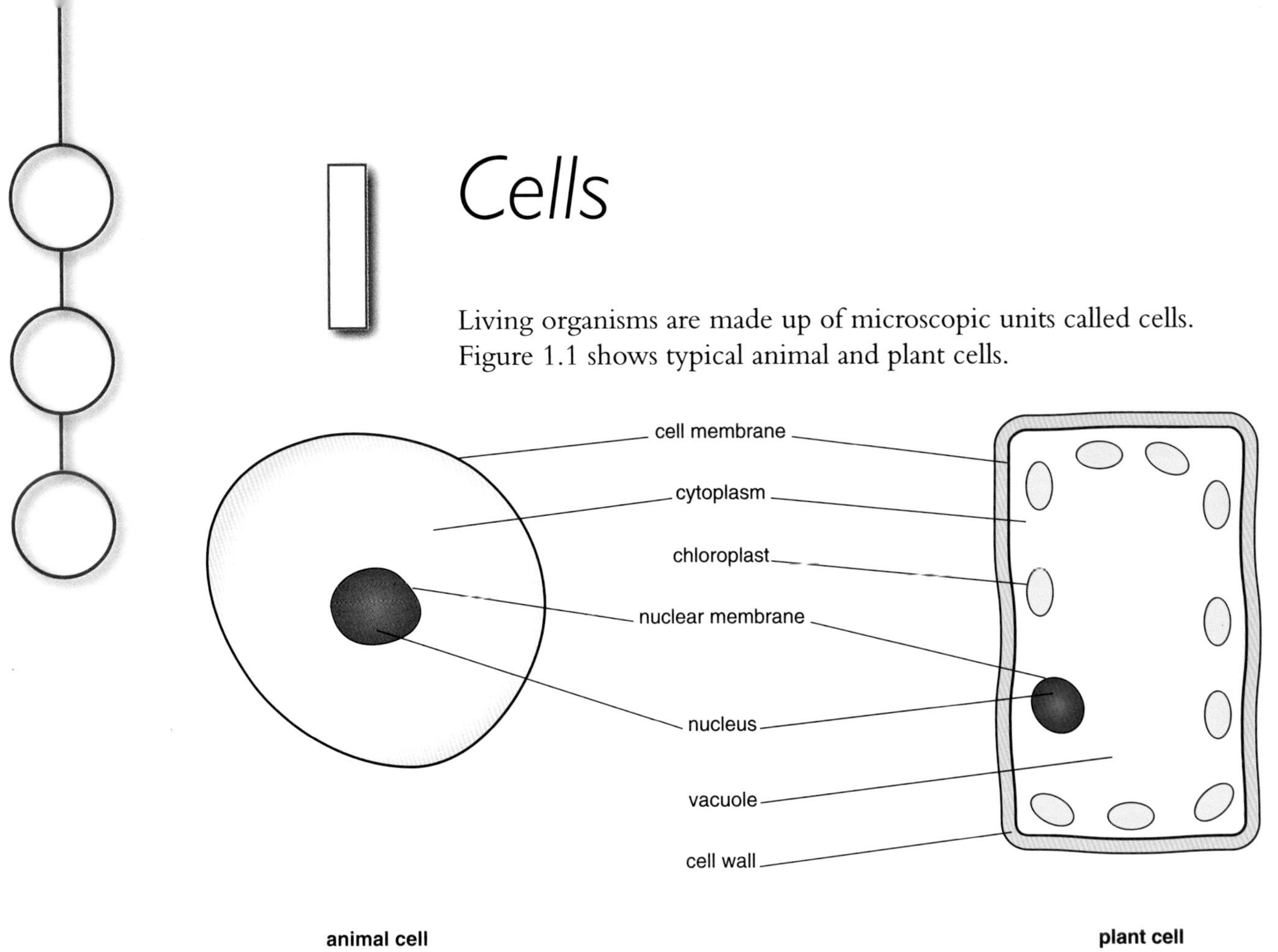

Figure 1.1 An animal cell and a plant cell

Animal cells

Animal cells contain a selectively permeable **cell membrane** that forms a boundary to the cell and controls what enters or leaves. The main part of the cell is the **cytoplasm** and this is where chemical reactions take place. The **nucleus** is the control centre of the cell and genetic information is contained in **chromosomes** inside it. The nucleus is surrounded by a **nuclear membrane**.

Plant cells

Like animal cells, plant cells have a cell membrane, cytoplasm and a nucleus. However, in addition they have:

* a **cellulose cell wall**, which is a rigid structure immediately outside the cell membrane that provides support
* a large permanent **vacuole** that contains cell sap and when full pushes the cell membrane against the cell wall, making the cell rigid and providing support
* **chloroplasts** that contain chlorophyll, which traps light and helps the plant make food during photosynthesis. Chloroplasts are not found in all plant cells – they are only found in green parts of the plant, particularly leaves.

Question

1 Draw a table to show the similarities and differences between plant and animal cells.

Observing cells using a microscope

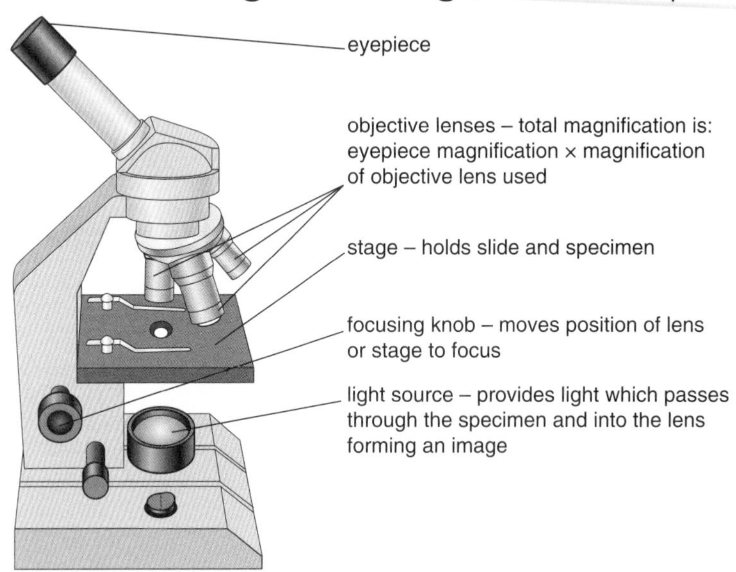

Figure 1.2 A light microscope and how it works

When using the microscope it is important to clip the slide containing the specimen you want to look at tightly into place on the stage. It is also important to focus at **low power** first. With low power you have a wider field of view (you can see more cells) and it is easier to find what you are looking for. It is also much easier to focus at low power.

After you observe the specimen at low power you may want to look at it at a higher magnification. Change the objective lens to give you a **higher power** and then refocus.

Note: when focusing it is very important that the objective lens doesn't come into contact with the slide – this is particularly likely to happen at high power. To prevent this you can move the lens down until it is *almost* touching the slide before you attempt to focus, and then focus as you move the slide and lens *further apart* rather than attempting to focus as the slide and lens are moving closer together.

Making slides

Most slides containing animal cells will already be prepared for you but you should get a chance to prepare slides containing plant cells. The procedure for making a slide containing onion cells is outlined below.

1. Peel a small piece of thin and transparent onion epidermis from the inside of an onion.
2. Using forceps place the onion epidermis evenly on a microscope slide.
3. Add water using a drop pipette to the onion epidermis to stop it drying out. (You could add a few drops of iodine instead of water – the iodine stains the cells, making certain parts such as the nucleus more obvious.)

4 Gently lower a coverslip onto the onion epidermis using a pointed needle or forceps. It is better to lower the coverslip one end first as this prevents trapping air bubbles. The coverslip will help protect the lens should the lens make contact with the slide and will prevent the cells drying out.

Note: Sometimes you will see black rings in your slide preparation – these are probably trapped air bubbles.

Calculating magnification using a scale bar

Microscopic and small objects are sometimes magnified before being represented as drawings or photographs in newspapers, books or even exam papers. If there is a scale bar available it is possible to work out the magnification used and the real size of the object that is magnified. In Figure 1.3 the scale bar represents a length of 0.1 mm *before* magnification. If you measure the scale bar you will find that it now measures 2 cm (20 mm). Therefore the onion cell has been magnified 200 times (20/0.1). If you measure the length of the onion cell on the page you will find that it measures 5 cm (50 mm); therefore its real length is 50 mm/200 = 0.25 mm.

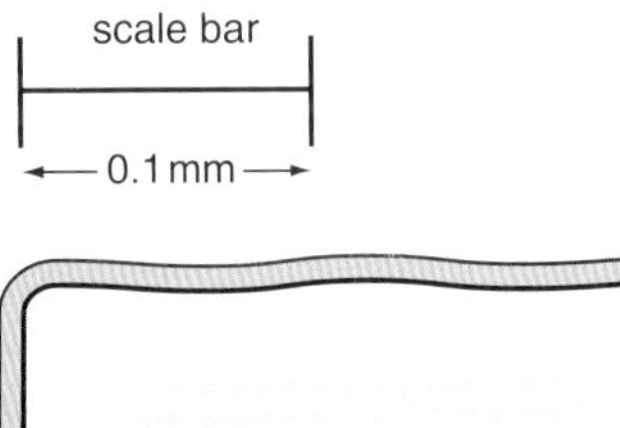

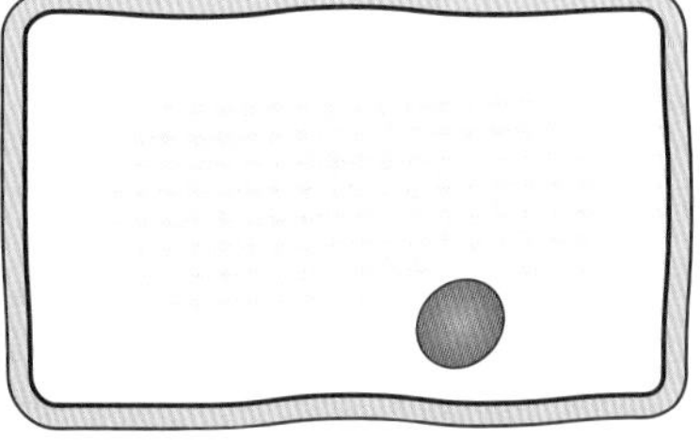

Figure 1.3 Measuring cell size

Foundation tier candidates will only be expected to calculate the size of a feature in a diagram or photograph, not the magnification involved. In Figure 1.3 you should be able to use the scale bar to work out that the onion cell is 0.25 mm in length – check and see how this is done.

► Bacterial cells

Bacteria are microscopic single-celled organisms (microorganisms). They are neither plant nor animal, largely because their cell structure is very different. For example, genetic material (DNA) is in the form of a loop – there are no chromosomes and no nucleus. Small circular rings of DNA, called **plasmids**, are also present. A cell wall is present but it is not made of cellulose. Figure 1.4 shows a typical bacterial cell.

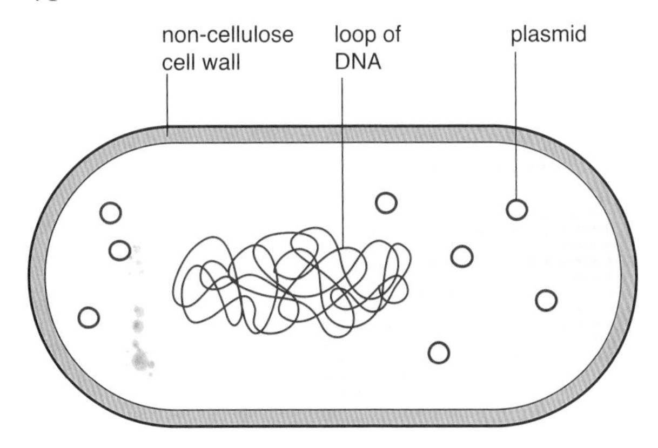

Figure 1.4 A typical bacterial cell

Question

2 State two things that plant and bacterial cells have in common and two differences between them.

The movement of substances into and out of cells

In a living cell it is important that essential materials can enter and waste products can leave. These substances must be able to pass through the plasma (cell) membrane. It is the cell membrane that controls what passes in or out. The membrane will allow some substances through but will prevent the movement of others – it is partially or **selectively permeable**.

Diffusion is the random movement of a substance (usually a simple, soluble molecule) from where it is in high concentration to where the concentration is lower. Diffusion is particularly important in the passage of gases through cell membranes. Examples include the diffusion of oxygen from the air spaces in the lungs into the bloodstream. The movement of gases into and out of the leaves during photosynthesis is another example of diffusion.

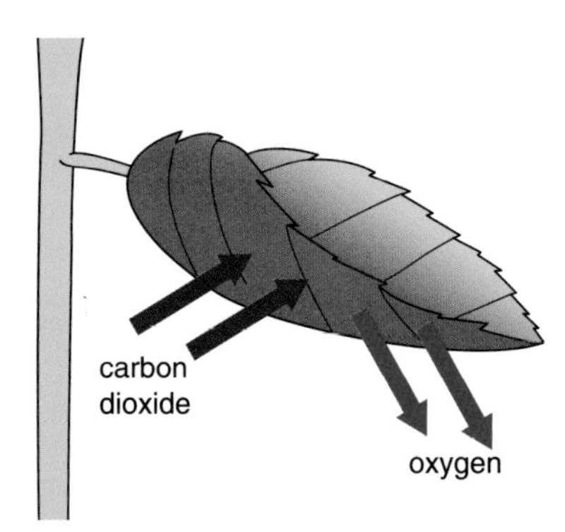

Figure 1.5 Diffusion is important in the movement of gases into and out of leaves as shown in photosynthesis

Tissues, organs, organ systems and organisms

It is important that the different cells in multi-celled organisms are not just randomly situated but are organised in an effective manner. There are different levels of organisation.

1 Similar cells grouped together are called a **tissue**. Examples of animal tissues include blood tissue and skin tissue.

2 An **organ** is a structure made of several types of tissue that carries out a particular function – the heart contains muscle, nerve and blood tissues and its function is to pump blood around the body.

3 Organs that operate together to carry out a particular function are grouped together as **organ systems**.

4 All the organ systems in a complex animal such as a human are grouped together to make the **organism**. The following table summarises three of the organ systems in humans.

Organ system	Function	Main organs include
Digestive	Breaking down of large food molecules into simple, soluble molecules that can be absorbed into the blood	Stomach, small intestine
Nervous	Responding to stimuli and making responses	Brain, spinal cord
Reproductive	Production of young	Testes, ovaries, uterus

Similar cells also group together to form tissues in plants. The leaf is an example of an organ in a plant.

The organisation of cells into tissues, organs and organ systems:

* improves exchange with the environment, e.g. lungs and gaseous exchange
* helps transport substances around the body, e.g. the circulatory system
* helps communication between cells, e.g. the nervous system and hormones.

► Stem cells

Stem cells are the very simple and undifferentiated cells that are found in young animal embryos (including human embryos) long before they become recognisable as a particular type of living organism. Unlike other cells, stem cells have a very unusual feature – they can **differentiate** into *any* of the **specialised cell types** found in the body. As an embryo develops the stem cells become differentiated into particular cell types and lose their ability to differentiate into any cell type.

Stem cell research is an important area of medical research. However, some people are concerned about it. They may have ethical issues with allowing embryos to develop purely for the purposes of stem cell research or they may worry about the idea of 'designer babies'. The Government carefully controls the limits of research in this area.

New research, including stem cell research, is validated (or rejected) by other scientists, expert in the subject area. This process of peer review involves a rigorous examination of new scientific advances.

Questions

3 a) Suggest one reason why stem cells are of interest to medical researchers.

b) Why is it important that government committees regulating stem cell research are supported by scientific advisers?

► Patterns of growth and development in plant and animal cells

Plants and animals grow according to very different patterns. **Animals** grow (fairly evenly) all over their body. This is why the shape of a baby is broadly similar to that of an adult. **Plants** grow in particular regions called **apices**. Shoot and root tips are examples of plant apices. At apices, growth and division take place, giving the characteristic **branching pattern** seen in plants.

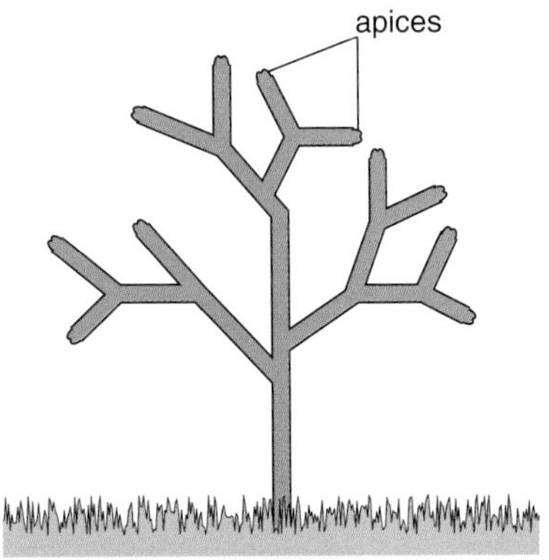

Figure 1.6 The typical branching pattern of plant growth

Question

4 Use Figure 1.6 to suggest why gardeners often cut off the upper tips of plant shoots to give bushier growth.

Exam questions

1 a) The diagram shows a microscope.

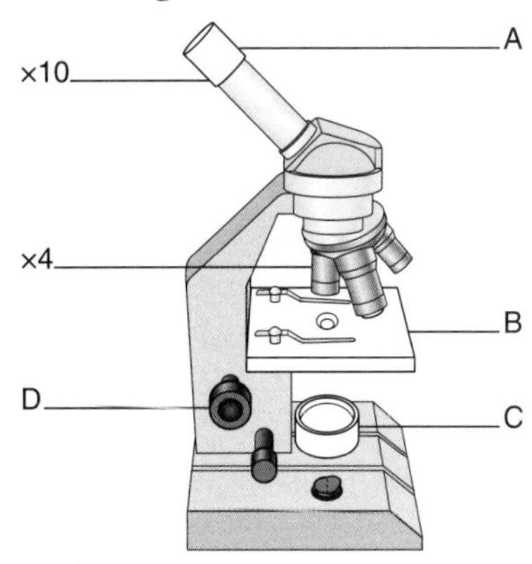

i) Name parts A, B and C. *(3 marks)*

ii) Give the function of part D when viewing the image. *(1 mark)*

iii) Calculate the magnification being used. Show your working. *(2 marks)*

b) The diagram shows how to prepare a slide to examine root cells.

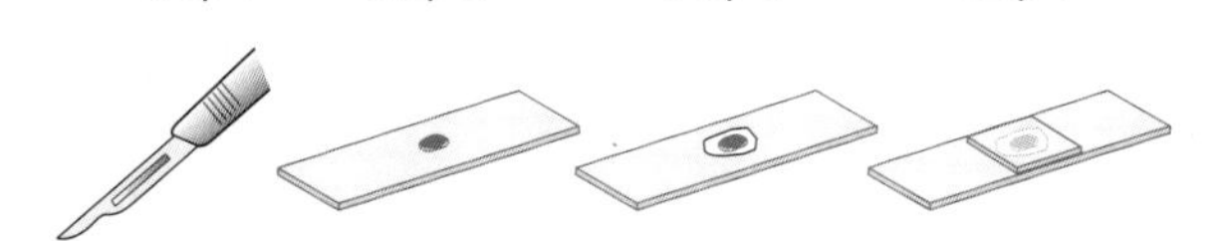

Step 1: Root cut into thin slices

Step 2: Thin slice placed on slide

Step 3: Dye added to slide

Step 4:

i) Suggest why dye was added to the slide in step 3. *(1 mark)*

ii) Use the diagram to describe what is being done at step 4. *(1 mark)*

c) The diagram shows a root hair cell viewed under high power.

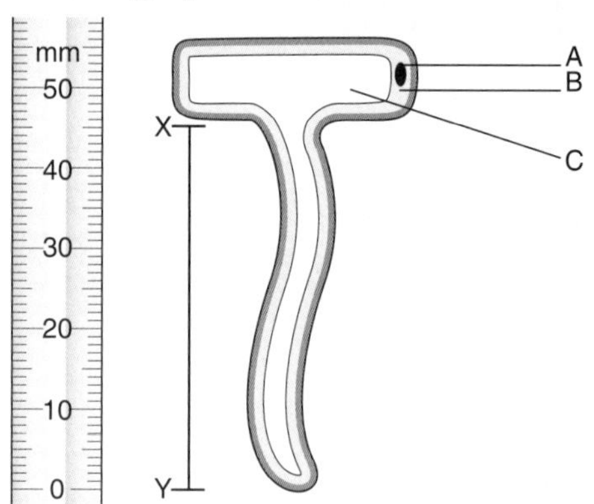

i) Name parts A, B and C. *(3 marks)*

ii) Use the scale to measure the length of the root hair from X to Y. *(1 mark)*

2 a) Copy and complete the magnified diagram of a leaf cell by drawing and labelling three structures found only in plant cells. *(6 marks)*

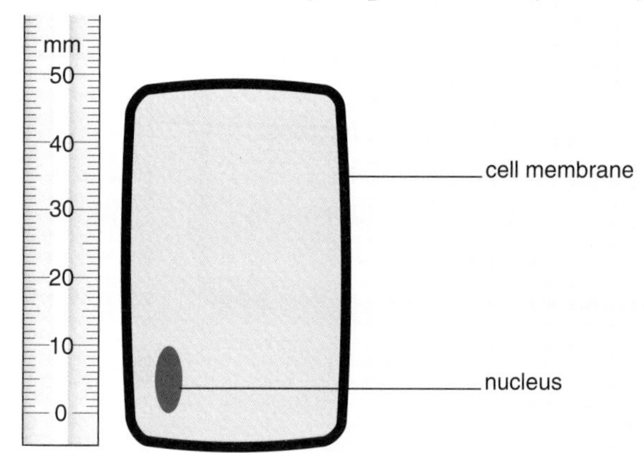

b) i) Use the scale to measure the diameter of the nucleus. *(1 mark)*

ii) This diagram has been magnified 100 times. Calculate the actual size of the nucleus in millimetres. Show your working. *(2 marks)*

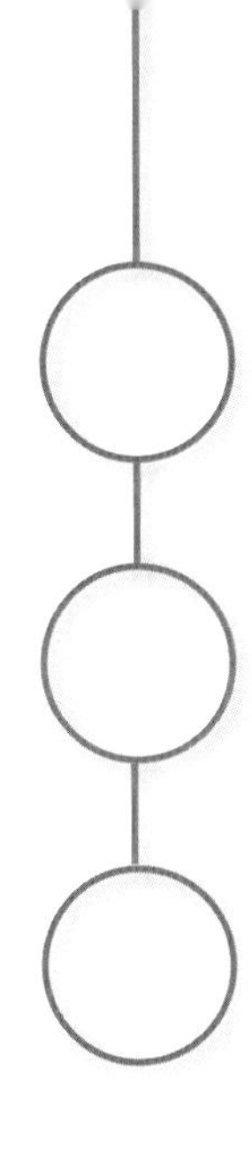

2 Photosynthesis and Plants

All living organisms require energy to survive. Energy comes from food. A very important difference between plants and animals is the source of their food. Animals must feed on material that is, or was, plant or animal. Plants are able to use light energy to make their own food.

Plants do this by converting the raw materials of carbon dioxide and water to glucose (sugar). The glucose is usually converted immediately into starch. This process is called **photosynthesis** and it takes place in the green parts of plants, particularly in the leaves. The green pigment **chlorophyll** is an important part of photosynthesis as it traps the light energy from the Sun that is needed to drive the process. Oxygen is produced as a waste product. Photosynthesis can be summarised by the equation:

$$\text{carbon dioxide} + \text{water} \xrightarrow[\text{by chlorophyll}]{\text{light energy trapped}} \text{glucose} + \text{oxygen}$$

Higher Tier GCSE Biology candidates need to know the balanced chemical equation for photosynthesis:

$$6CO_2 + 6H_2O \rightarrow C_6H_{12}O_6 + 6O_2$$

In photosynthesis, light energy from the Sun is converted into chemical energy (food). Photosynthesis is important for animals, as well as plants, as it provides a source of food and also releases oxygen back into the atmosphere.

Leaves (and other parts of the plant) that are carrying out photosyntheses in bright light will take carbon dioxide into the leaves and oxygen will move out. Not suprisingly, the brighter the light the faster the process will take place.

Photosynthesis experiments

It is possible to carry out investigations to show that photosynthesis is taking place or that particular raw materials are needed for the process.

The starch test

This test can be used to show that starch is produced in green leaves during photosynthesis. The starch test consists of a series of steps. These are:

* A leaf is removed from a plant that has been placed in bright light.
* The leaf is placed in boiling water for at least 30 seconds. This will kill the leaf and ensure that no further reactions can take place.

* The leaf is then placed in boiling ethanol (Figure 2.1a). This will remove the chlorophyll from the leaf. This procedure must take place using a water bath as ethanol is flammable and must not be exposed to a direct flame.
* The leaf should then be dipped in boiling water again. This will make the leaf soft again as the ethanol makes the leaf very brittle.
* The leaf is then spread out on a white tile and iodine is added to the leaf (Figure 2.1b).
* If starch is present the iodine will turn the starch blue-black. If there is no starch present the leaf will remain a yellow-brown colour (the colour of the iodine).

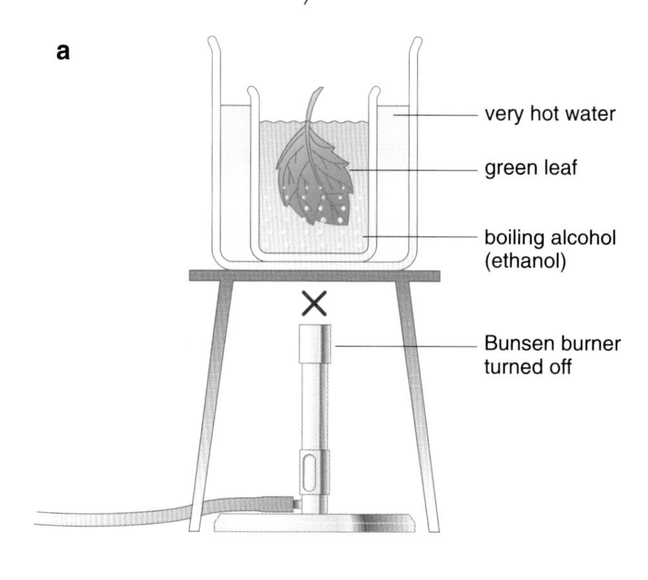

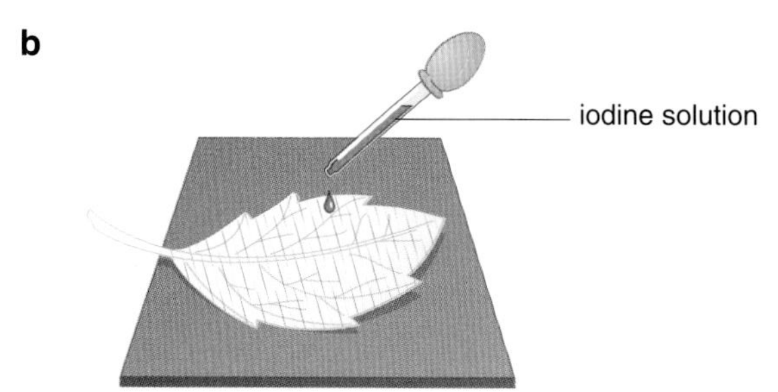

Figure 2.1 Testing for starch

Investigations showing that light, carbon dioxide and chlorophyll are necessary for photosynthesis

To carry out these experiments it is necessary to **destarch** the leaves of the plant first. Leaves can be destarched by leaving the plant in the dark for at least 2 days. This will ensure that any starch already in the leaves will be removed and stored elsewhere in the plant, or used by the plant during this period. The importance of this is that if the starch test at the end of the investigation is positive, it shows that the starch must have been produced during the period of the investigation.

Question

1 If light is essential for photosynthesis what results would you expect?

Light

A leaf is partially covered with black paper or light-proof foil as shown in Figure 2.2. After a period of time the leaf is tested for starch as described above.

Carbon dioxide

Question

2 If carbon dioxide is an essential raw material for photosynthesis what results would you expect?

To show that carbon dioxide is an essential raw material for photosynthesis it is necessary to compare a leaf that is deprived of carbon dioxide with a leaf that has a good supply of carbon dioxide. This can be achieved by preparing two leaves as shown in Figure 2.2. Sodium hydroxide will remove the carbon dioxide from the air surrounding the experimental leaf. The control leaf will only have water (or a chemical that increases carbon dioxide levels) in its flask and therefore there will be carbon dioxide present.

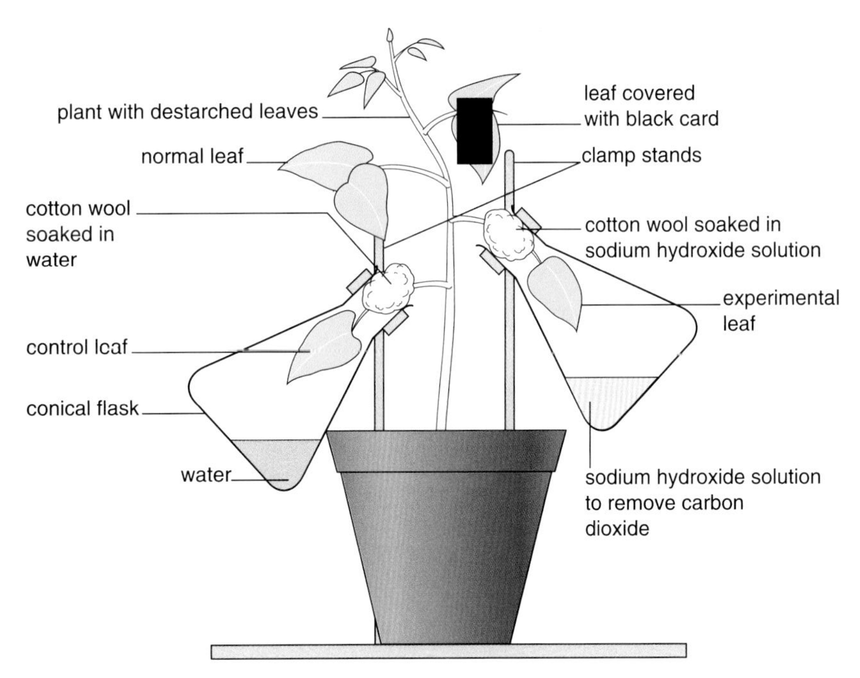

Figure 2.2 Investigating the raw materials for photosynthesis

Chlorophyll

Some plants have leaves that are part green and part white. These leaves are described as being **variegated**. If a variegated leaf is tested for starch it will be apparent that starch is only produced in the green parts of the leaves. This shows that chlorophyll, the substance that gives leaves their green colour, is necessary for photosynthesis.

Measuring the rate of photosynthesis in different conditions

Using apparatus similar to that shown in Figure 2.3 it is possible to demonstrate that oxygen is produced in photosynthesis. The rate of photosynthesis will affect the rate at which the bubbles of oxygen will be given off and this can be used to compare photosynthesis rates in different conditions. For example, by moving the position of the lamp it is possible to investigate the effect of light intensity on photosynthesis.

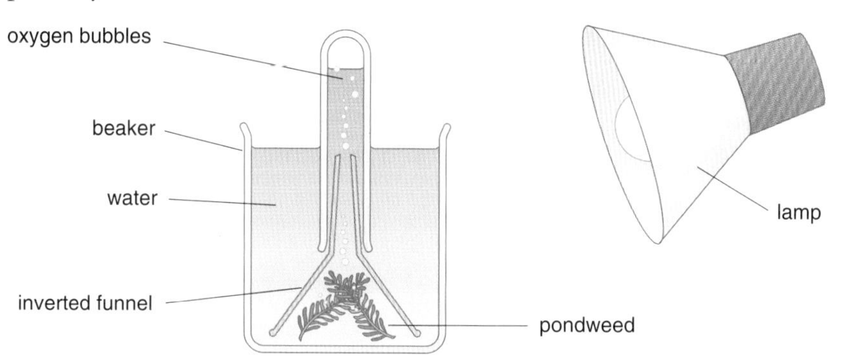

Figure 2.3 Measuring the rate of photosynthesis

Question

3 Why is it more accurate to measure the volume of oxygen collected?

The rate of photosynthesis can be more accurately calculated by measuring the volume of oxygen produced.

Alternatively, an oxygen electrode connected to a data logger can be used to measure changes in oxygen levels.

Have a look at some photosynthesis experiments on the internet.

Search

- photosynthesis experiments

The leaf – the site of photosynthesis

In most plants the process of photosynthesis takes place in the leaves. Leaves come in many shapes and sizes but to allow photosynthesis to take place efficiently they are usually highly adapted for:

* light absorption
* gas exchange.

The way in which leaves are arranged on a plant ensures that each leaf can absorb as much light as possible and that as far as possible each leaf is not in the shade of other leaves. The section through a leaf shown in Figure 2.4 shows many other ways in which a leaf is designed to aid light absorption and encourage gas exchange.

Use a microscope to investigate a cross section of a mesophytic (typical unspecialised) leaf. The leaf you examine may be different from the one described in this book, but you should look for as many adaptations for maximising light absorption and gas exchange as possible.

Light absorption in a leaf is maximised by:

* the short distance from top to bottom, which allows all the cells to receive light
* the large surface area
* the thin transparent **cuticle** that covers the epidermis (single cell layer covering leaf), which reduces water loss by evaporation, but does not prevent light entering the leaf
* the presence of chloroplasts rich in the pigment chlorophyll that absorbs the light
* the regular structure of the **palisade mesophyll**, which ensures that many cells rich in chloroplasts are packed together near the upper surface of the leaf.

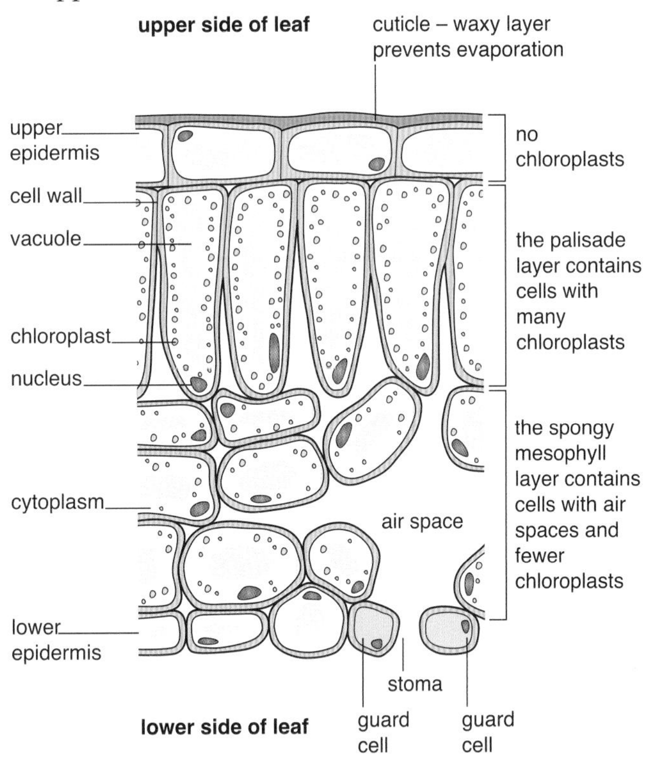

Figure 2.4 Cross-section of a mesophytic leaf

Gas exchange in a leaf is maximised by:

* the intercellular spaces in the spongy mesophyll, which allow carbon dioxide to enter, and oxygen to leave the photosynthesising cells that are mainly concentrated in the palisade layer
* stomata, which allow carbon dioxide and oxygen to enter and leave the leaf. Stomata are small pores that can occur between cells in the epidermis on the bottom surfaces of leaves. Each stoma is surrounded by two guard cells that regulate the opening and closing of stomata. In many plants the stomata are open during the day and closed at night.

Question

4 Why do some plants, such as water lilies, only have stomata on their upper leaf surfaces?

In some plants stomata can occur on both the upper and the lower leaf surface. Some plants have all their stomata on their upper leaf surfaces.

Using the products of photosynthesis

The glucose that is produced in photosynthesis can be used in a number of ways or converted into a range of products that the plant requires. These include:

* **respiration** – the glucose is used in respiration to provide energy
* **storage** – in many plants the glucose is converted into starch and oils for storage
* **useful substances** – the glucose can be converted into a range of useful products including cellulose (for cell walls), chlorophyll and protein for growth.

Factors affecting the rate of photosynthesis

The rate at which photosynthesis occurs depends on the availability of the raw materials needed for the process. Levels of carbon dioxide and light will directly affect the rate of photosynthesis, as will the availability of water to some extent. As temperature affects the rate of all reactions it will influence the rate at which photosynthesis takes place.

When photosynthesis is taking place at its maximum rate all of these environmental factors must be present at peak or optimum levels. However, if one (or more) factor is in short supply the rate of photosynthesis will be limited. These raw materials become **limiting factors** and the rate of photosynthesis will be determined by whichever factor is in shortest supply.

Figure 2.5 shows the effect of light intensity on the rate of photosynthesis. Part (a) shows how temperature (cold and hot days) can further influence the rate.

As light intensity increases, irrespective of temperature, the rate of photosynthesis increases up to a point where the graph begins to level off and form a plateau. As an increase in light intensity causes an increase in photosynthesis at the lower light levels, the amount of light must be limiting the rate at which photosynthesis occurs. Within the plateau part of the graph, further increases in light intensity do not lead to an increase in photosynthesis, therefore something else must be limiting the rate.

The effect of temperature can be explained by comparing the rates of photosynthesis on the cold and hot days. On the hot day photosynthesis occurs at a higher rate at higher light intensities when compared to the cooler day. Therefore we can conclude that temperature is limiting the rate at the higher light intensities on the cooler day.

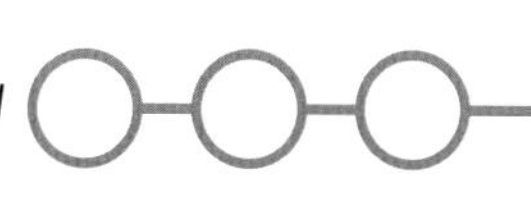

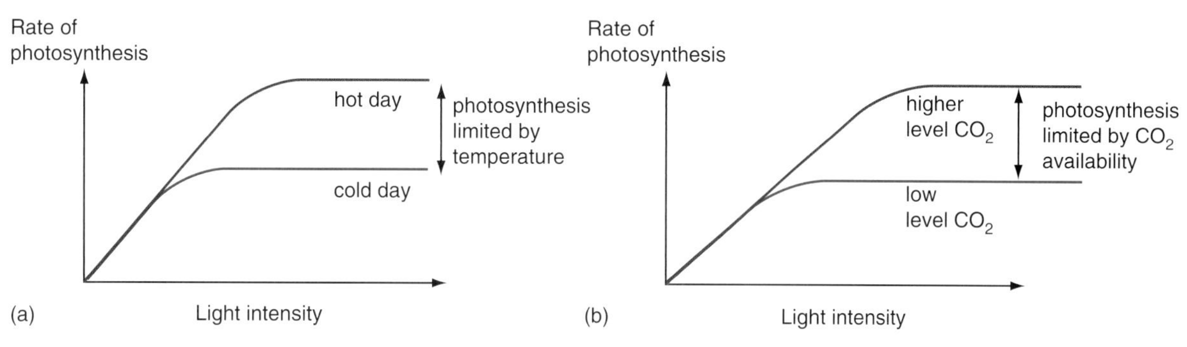

Figure 2.5 These graphs show how the rate of photosynthesis is affected by increasing light intensity on (a) a hot and a cold day; (b) higher and lower CO_2 levels

> **Question**
>
> 5 The graphs in Figure 2.5 show that the raw materials for photosynthesis interact with each other and that while some of the essential materials can be present in abundance it only takes one to be in short supply to limit the rate. Which of the factors discussed above do you think might limit the rate of photosynthesis in the following situations:
>
> a) during a bright winter afternoon in a British grassland
>
> b) in a cornfield in mid-summer sunshine in Southern France?

It is possible that the rate of photosynthesis may still not be at its maximum where the rate has plateaued on the hot day at the highest light intensities. It is possible that carbon dioxide could be a limiting factor in these conditions. To test this we would need to increase carbon dioxide levels to see if this has any effect.

Figure 2.5b shows how light intensity affects the rate of photosynthesis at different carbon dioxide levels.

Again, light intensity is limiting the rate of photosynthesis at low light levels. The fact that an increased carbon dioxide level leads to a higher rate of photosynthesis at higher light intensities shows that the low carbon dioxide level was a limiting factor once light levels had ceased to be limiting.

The economic implications of enhancing environmental factors in crop production

To maximise photosynthesis (and profit) it is important to ensure that all the factors that affect the rate of photosynthesis are kept at their optimum levels, for example by using artificial light, carbon dioxide enrichment and fertiliser.

Figure 2.6 shows how the environment in a glasshouse can be controlled to maximise crop production.

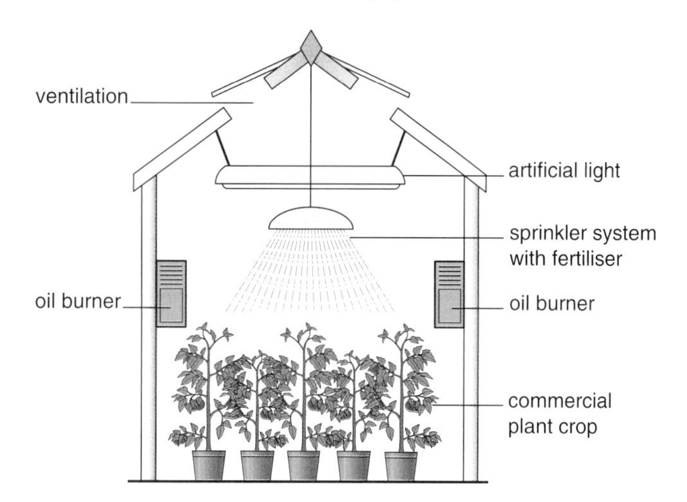

Figure 2.6 Maximising photosynthesis

However, to obtain maximum profit it is important that none of these methods brings the environmental factor *above* its optimum level. Additionally, the increased profit gained by enhancing environmental factors must be greater than the extra costs of adding additional fertiliser, carbon dioxide and light.

The balance between photosynthesis and respiration

All living organisms respire. In plant respiration the glucose produced in photosynthesis is broken down to release energy. Plants require oxygen to respire and they produce carbon dioxide as a waste product. These gases enter the leaves through the stomata. During the night when there is no light for photosynthesis, respiration will be the only process involving gas exchange that takes place. Therefore oxygen will enter the leaf and carbon dioxide will leave. However, during the day when photosynthesis is occurring both processes will take place. When the light intensity is high the rate of photosynthesis will exceed the rate of respiration. When this happens carbon dioxide enters the leaves and oxygen moves out. There will be times during the day when the light intensity is low, causing photosynthesis to take place very slowly. At these points, usually at dawn and dusk, the rates of respiration and photosynthesis are equal and there will be no overall, or net, gas exchange. This point is called the **compensation point**.

The movement of carbon dioxide and oxygen into and out of plants can be determined using hydrogencarbonate indicator. Hydrogencarbonate indicator is bright red in normal atmospheric carbon dioxide levels. If there is an increase in carbon dioxide levels the indicator will change colour to yellow. A decrease in carbon dioxide levels will turn the indicator purple. Figure 2.7 shows how the indicator can be used to show gas exchange in living organisms.

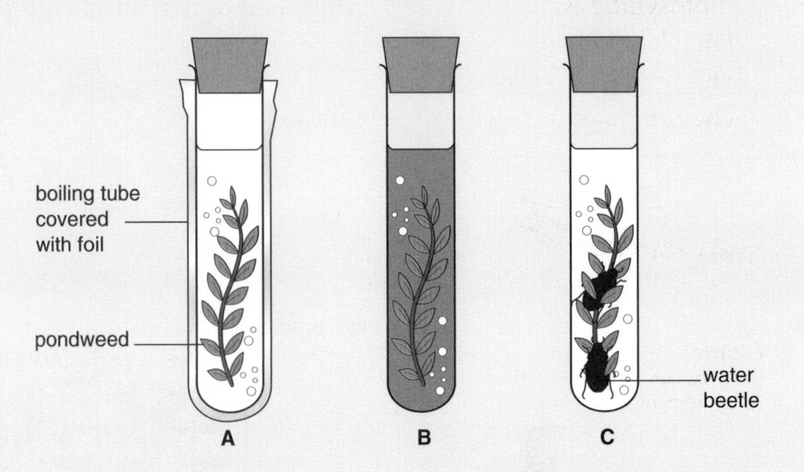

Figure 2.7 The effect of photosynthesis and respiration on gas exchange. Each boiling tube was filled with hydrogencarbonate indicator and placed in bright light for 1 hour

The results are explained in the following table.

Tube	Colour at start	Colour at end	Reason for change
A	Red	Yellow	The foil strip stops light entering and photosynthesis does not occur. Respiration increases carbon dioxide levels.
B	Red	Purple	Both photosynthesis and respiration are taking place in the pondweed. As the rate of photosynthesis is faster than the rate of respiration more carbon dioxide enters the plant than is produced.
C	Red	Yellow	The water beetles produce more carbon dioxide in respiration than the pondweed takes in for photosynthesis.

Questions

6 Look at Figure 2.7 and the table above. Suggest what colour change you would expect if the following changes were made. Explain your answers in each case.

a) There was only partial shading in tube A allowing some light to enter.

b) More pondweed was introduced into tube C.

Exam questions

1 a) Photosynthesis occurs in plant leaves.

i) Name the chemical in leaves that absorbs light for photosynthesis. *(1 mark)*

ii) Copy and complete the word equation for photosynthesis.

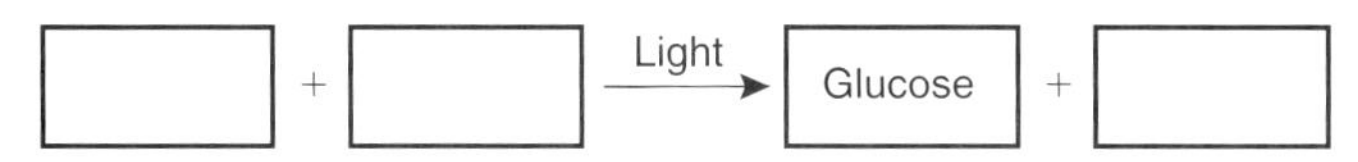

b) The diagram shows the apparatus used to investigate if carbon dioxide is needed for photosynthesis.
The plant was destarched and then the leaves were sealed in glass flasks.
The plant was then left in sunlight for 12 hours.

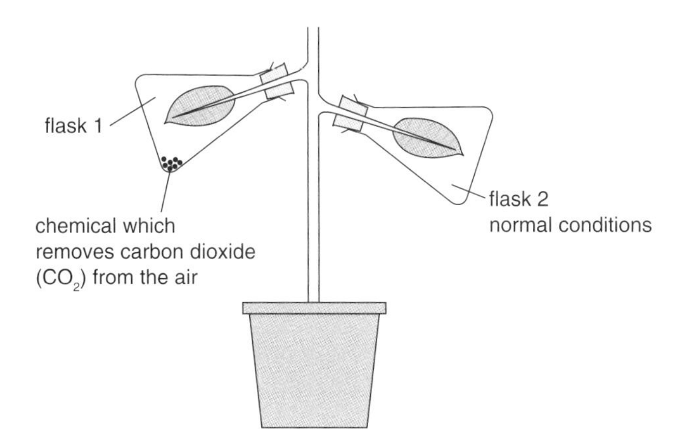

i) How was the plant destarched? *(1 mark)*

The diagrams show the stages in the starch test on the leaves from the plant.

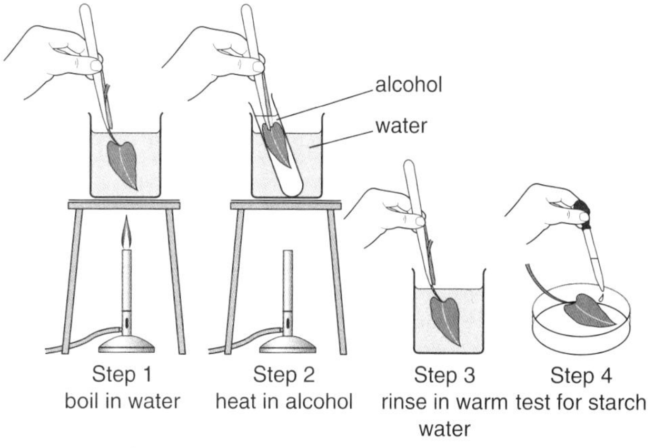

ii) What is the purpose of:
Step 2
Step 3? *(2 marks)*

iii) Why is the Bunsen burner turned off before Step 2? *(1 mark)*

iv) In Step 4, iodine solution is added to the leaves to test for starch.
What colour would you expect to obtain when iodine is added to the leaf from:
Flask 1
Flask 2? *(2 marks)*

2 When placed in sunlight for 8 hours, the apparatus below can be used to investigate one of the requirements for photosynthesis.

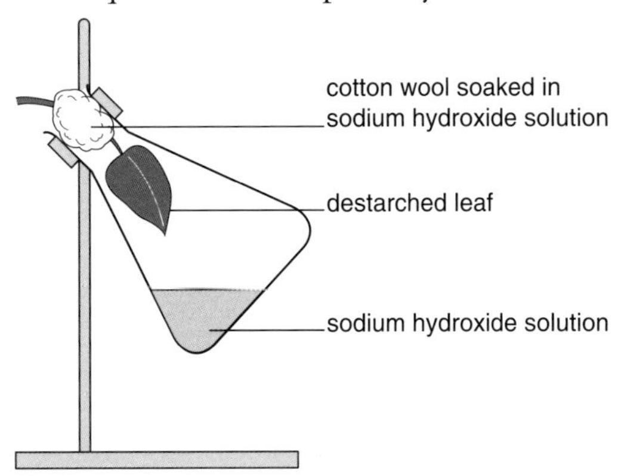

a) Describe and explain the results of a starch test on the leaf after 8 hours. *(3 marks)*

b) Describe how a leaf can be tested for starch. *(4 marks)*

3 The diagram shows the apparatus used to investigate photosynthesis by recording the number of bubbles of oxygen produced by pondweed.

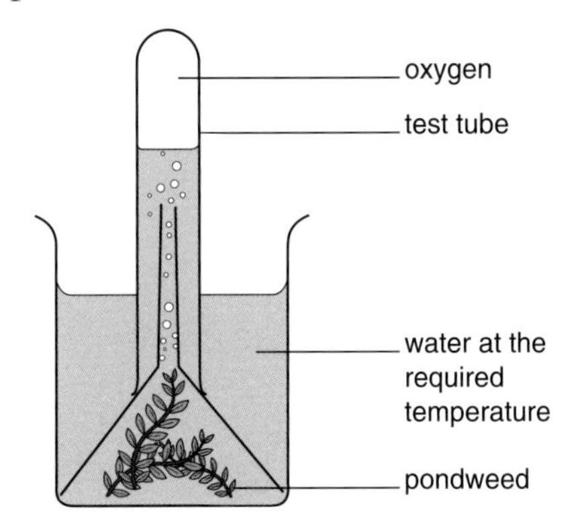

The table records the number of bubbles counted in 2 minute intervals at different temperatures. An interval of 5 minutes was left at each new temperature before bubbles were counted.

Temperature/ °C	5	10	15	20	25	30	35	40
Number of bubbles produced in 2 minutes	3	9	15	22	27	31	31	31

a) Draw a graph of these results, showing the number of bubbles of oxygen produced in 2 minutes against the temperature of the water/°C. *(4 marks)*

b) Why was a 5-minute interval left at each new temperature before the bubbles were counted? *(1 mark)*

c) Using evidence from the graph, state the optimum temperature for photosynthesis in pondweed. *(1 mark)*

d) If room temperature is 18 °C use your graph to estimate the number of bubbles produced in a 2-minute period at room temperature. *(1 mark)*

e) State **one** other factor that could affect the rate of photosynthesis in pondweed and suggest how this apparatus could be changed to investigate this. *(2 marks)*

f) How could you set up an experiment to show that the oxygen is coming from the plant and not from the water? *(1 mark)*

4 A market gardener investigated how the rate of photosynthesis was affected by increasing the concentration of carbon dioxide in the surrounding air. The results are shown below.

Carbon dioxide concentration/% in air	Rate of photosynthesis (arbitrary units) in high light intensity
0	0
0.02	18
0.04	
0.06	54
0.08	72
0.10	86
0.12	94
0.14	95
0.16	95
0.18	95
0.20	95

a) Use these results to draw a line graph to show how the rate of photosynthesis changes with increasing carbon dioxide concentration. *(1 mark)*

b) The concentration of carbon dioxide in normal air is 0.04%. When doing the experiment the market gardener forgot to record the rate of photosynthesis at 0.04% carbon dioxide concentration.
From your graph suggest what the rate of photosynthesis would be at this carbon dioxide concentration. *(1 mark)*

c) Draw another line on the graph showing the results you would expect to get if the investigation had been carried out at a lower light intensity. *(2 marks)*

d) Market gardeners sometimes add carbon dioxide to their greenhouses. What is the advantage of this to the gardener? *(1 mark)*

e) i) What is meant by the term limiting factor? *(1 mark)*

ii) Apart from carbon dioxide and the light, name **one** other limiting factor. *(1 mark)*

5 The diagram shows part of a leaf.

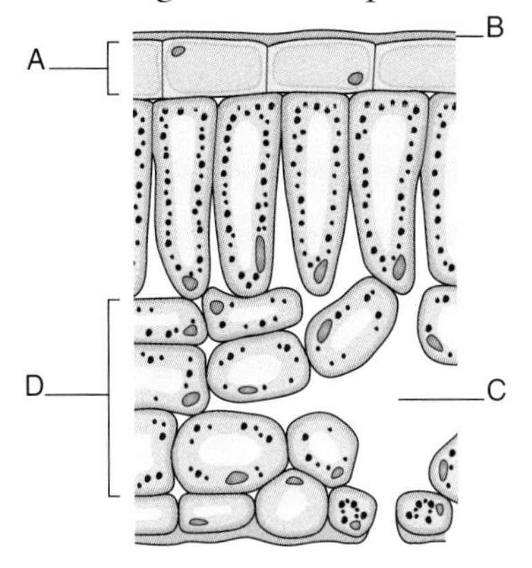

a) i) Name parts A and D. *(2 marks)*

ii) Describe how part B is adapted to its function. *(2 marks)*

iii) Describe the role of part C. *(2 marks)*

iv) Name the type of cell that carries out most photosynthesis. *(1 mark)*

v) Name the type of cell that controls the movement of carbon dioxide in and out of the leaf. *(1 mark)*

b) The graph shows the effect of increasing the light intensity on the rate of oxygen production, which is a method for measuring the rate of photosynthesis.

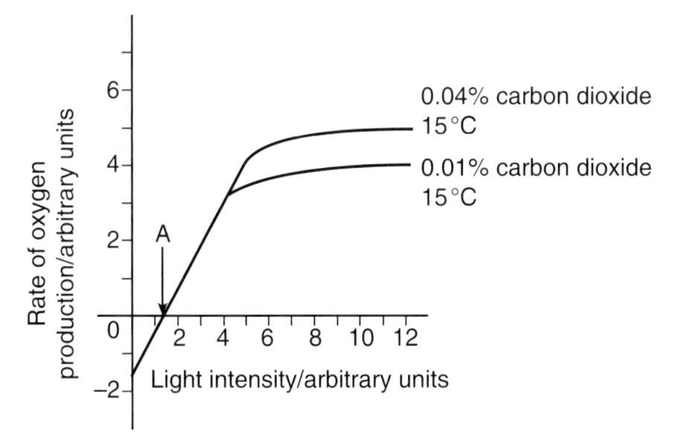

i) Sketch the graph and draw a curve showing the rate of photosynthesis at 0.04% carbon dioxide and 25 °C. *(2 marks)*

ii) Explain why market gardeners enrich the air in their greenhouses with carbon dioxide. *(2 marks)*

iii) Explain why the oxygen production is zero at point A. *(3 marks)*

6 The apparatus shown below was used to investigate the effect of light on gas exchange in a water plant. Hydrogencarbonate indicator was used to show any changes in carbon dioxide level. At normal levels of atmospheric carbon dioxide the indicator is red.

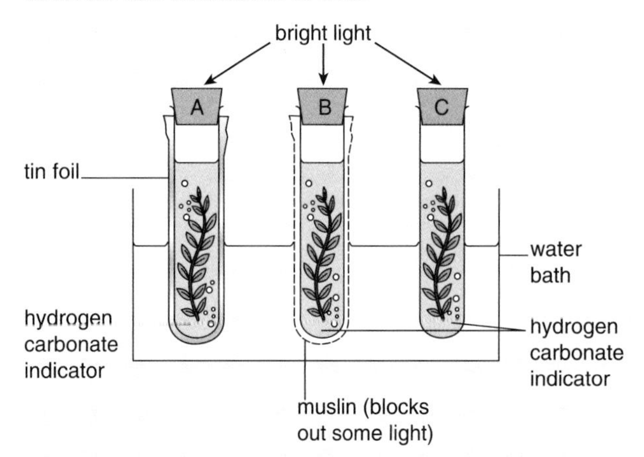

The experiment was left for 1 hour. Copy and complete the table below to show the results of the experiment and give an explanation of these results. *(5 marks)*

Tube	Initial colour of indicator	Final colour of indicator	Biological explanation of result
A	red		
B	red	red	
C	red		

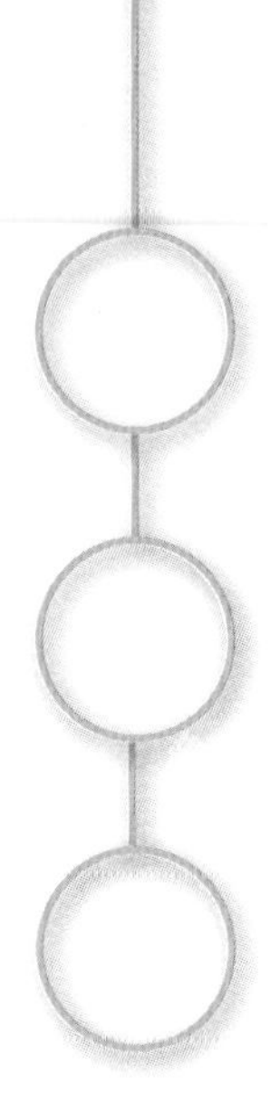

3 Nutrition and Health

Investigating the energy content of food

The apparatus in Figure 3.1 can be used to compare energy values in different foods such as crisps, bread and pasta. The food is held on the end of a pointed needle, ignited and then placed immediately underneath the test tube. The rise in temperature of the water in the test tube will give an indication of how much energy there is in the food.

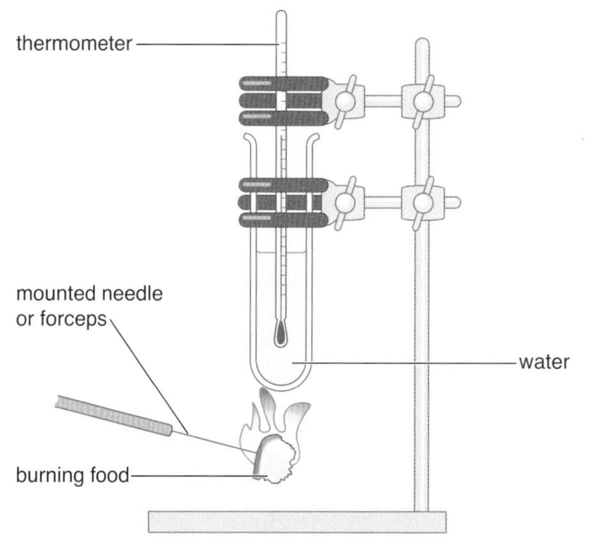

Figure 3.1 Measuring the energy content of food

Question

1 a) List three of the factors (variables) you would have to keep the same each time you test a food in the above experiment to give valid results (a fair test).
b) Suggest why the energy value you get from burning a particular food is likely to be an underestimation of the energy in the food sample.

Note: you can calculate how much energy is in each food sample by using the following equation.

energy released in joules (J) = mass of water (g) × rise in temp (°C) × 4.2

However, this equation is not necessary if you only want to *compare* the energy in different foods.

Although the energy values of foods can be compared in the laboratory as described above, food labels on packaging can also be used to compare energy values. It is also very useful to use food packaging information to check how accurate your comparisons in the laboratory are.

How much energy do we need?

Although different foods can provide us with different amounts of energy, it is also true that each of us requires different amounts of energy to keep us alive. Each person also requires different amounts of energy at different times of their life and depending on how active they are. Figure 3.2 shows how the energy requirement varies in some different situations.

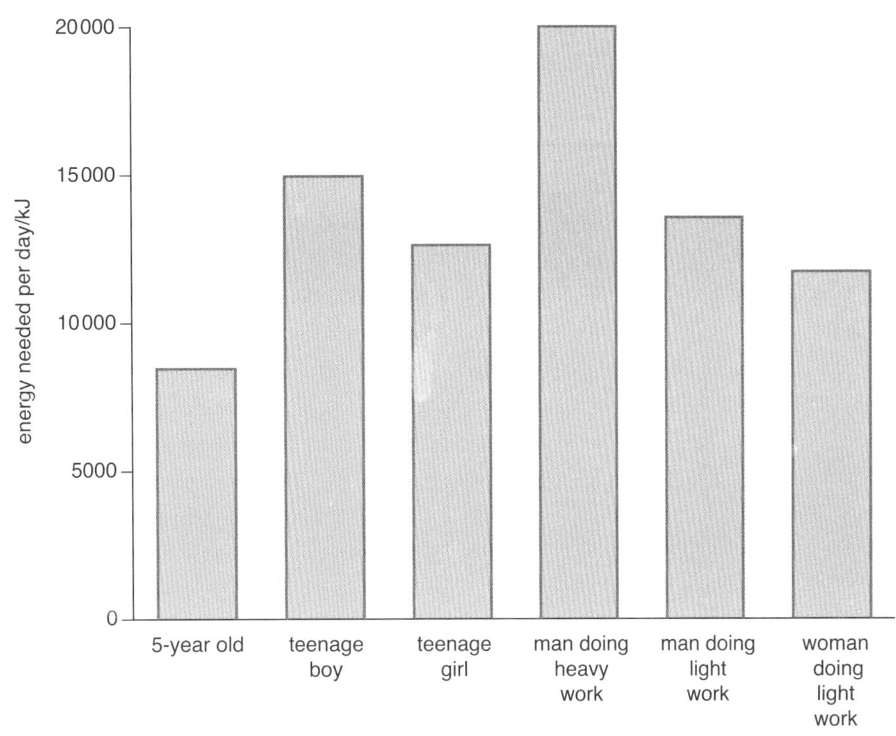

Figure 3.2 Energy needed per day by different people in different situations

You will be able to deduce from Figure 3.2 that there are three main factors that affect our energy requirement. These are:

* age
* gender (male or female)
* activity.

Of course, these factors often overlap. A very young baby will not be as active as a teenager and some men may need more energy than women because they have more physically demanding jobs. **Pregnant women** often have higher energy needs due to the high rates of growth in the developing foetus.

Obesity is caused by having a higher energy intake than the energy used (in exercise).

Calculating your Body Mass Index (BMI)

Health professionals often check patients' BMI – this information gives an indication of whether people are the right weight for their height.

To calculate BMI divide weight (kg) by the square of the height (m).

The following table gives some information about BMI values.

BMI Value	Description
Under 18.5	Underweight
18.5–24.9	Normal
25–29.9	Overweight
30+	Obese

Search
▶ body mass index

The internet contains a number of sites where if you type in height and weight the BMI will automatically be calculated.
Search using 'Body mass index' and try it!

Calculating your basal metabolic rate (BMR)

The BMR is a measure of minimum energy expenditure – this will occur when someone is inactive, resting in a reclining position. He/she must not be stressed due to hunger but it must be long enough after a meal for digestion to be complete.

Carry out an internet search for 'basal metabolic rate' or BMR and you will find calculators to help estimate BMR. Normally you have to add information such as height, weight, age and gender, as these affect BMR.

You can use the **Harris Benedict** equation to calculate how much energy is needed, in calories, taking account of approximate energy level as indicated in the table below.

Activity factor	Category
1.2	Minimal exercise
1.375	Lightly active
1.55	Moderately active
1.725	Very active
1.9	Extremely active

In each case multiply the BMR by the activity factor and this will tell you the number of calories needed to maintain current weight, i.e. not gaining or losing any weight. It is important to note that this value is only an estimate.

Question

2 a) State two body functions that we need energy for as part of our BMR.
b) Suggest why our BMR tends to decrease as we get older.

The link between health and diet – are we what we eat?

Our diet has a major effect on our health. There are a number of features of our diet that contribute to many people not being as healthy as they could be.

Too many of us do not make **healthy food choices**. This may be due to eating too many foods that are rich in **sugar**, **salt** and **fat**, or it may be not getting enough of other vital ingredients. Too many of us do not eat enough **fruit** and **vegetables**. It is very important to try to have the minimum 'five portions of fruit and vegetables' each day to ensure we get enough minerals, vitamins and fibre. Making healthy food choices really means ensuring that we try to have a **balanced diet**. While many students reading this book will know what healthy food choices are, how many of us actually make sure we eat a healthy diet?

Poor diet contributes to many health problems.

1 **Obesity** (being very overweight) is caused by taking in too much energy (sugar, starch or fat).

2 **Heart disease** is caused by cholesterol and other fatty substances being present in such high levels that they build up in the walls of the arteries. Over time this leads to a narrowing of the arteries, making it more difficult for blood to flow through them, and a heart attack may result.

3 **Strokes** – poor diet can damage the circulatory system in other parts of the body as well as the heart. If the build up of fat and subsequent blockage is in the **brain**, a stroke can result.

4 **High blood pressure** – obese people often have high blood pressure; the heart works harder to pump the blood round the extra tissues in the body. This can harm the heart itself due to greater wear and tear or can damage the blood vessels. Poor diet can also lead to high blood pressure due to narrowing of the arteries, making it harder to pump blood. Too much salt in the diet is another factor that contributes to high blood pressure.

5 **Diabetes** – a poorly balanced diet or obesity can lead to Type 2 diabetes. In Type 2 diabetes the body suffers sugar 'overload' and insulin stops being effective in controlling the sugar levels.

6 **Arthritis** – this is caused by damage to the joints, particularly in the knees, hips and lower back, due to the greater wear and tear these joints experience in someone who is very overweight.

Obesity – the costs to society

Over 20% of people in the UK are obese and the percentage is rapidly rising, having more than doubled over the last 20 years. The rising number of obese people significantly adds to the costs of the NHS, partially through treating people for the obesity itself, but more importantly through the costs of treating the medical conditions associated with obesity such as Type 2 diabetes and heart disease.

However, it is important to note that **inherited factors** also affect human health. For example, our cholesterol levels and the likelihood of suffering from heart disease are dependent on our genetics as well as our diet.

Investigating food samples using food tests

You should carry out food tests for sugars, starch, protein and fats on a range of foods. It is often necessary to break the food up using a pestle and mortar and to add a small quantity of water to make it into a solution before carrying out the test.

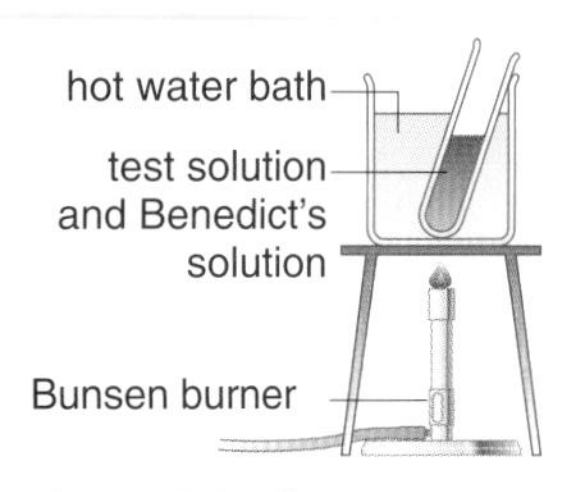

Figure 3.3a Sugar test

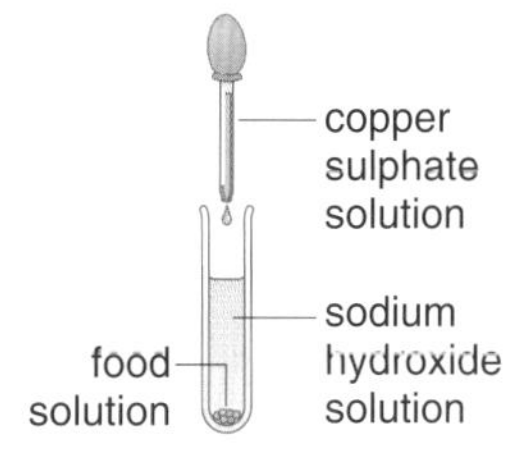

Figure 3.3b Protein test

Question

3 Plan an experiment to compare the amount of sugar in onion and in white chocolate. In your plan you should:
* state what you would do to make your results valid (a fair test)
* suggest why it is better to use white (and not brown) chocolate.

The following table outlines the tests you need to know.

Food	Name of test	Method	Positive result
Starch	Starch Test	Add iodine to the food	The iodine turns from **yellow-brown** to **blue-black**
Sugar	Benedict's Test	Add Benedict's solution to the food and heat in a water bath	Turns from **blue** to green then orange to **brick red precipitate** depending on how much sugar is present
Protein	Biuret Test	Add sodium hydroxide to the food solution, then add copper sulphate and shake	Turns from **blue** to **purple**
Fats	Ethanol	Mix the fat with ethanol	Turns from **clear** to form a **white emulsion**

Another food test – using DCPIP to test vitamin C content

If vitamin C is added to DCPIP the DCPIP will change colour from **blue** through pink to **colourless**.

The vitamin C content of a range of juices can be compared by recording how many drops of juice are required to turn the DCPIP colourless. The relative concentration of the juices (including natural, processed and boiled juices) can be calculated by comparing with a standard solution of ascorbic acid (vitamin C). Unlike most of the food tests described above, this test is **quantitative** in that specific values can be obtained for different juices.

The balanced diet

Irrespective of our age, gender or activity, to stay healthy we all need a balanced diet. People who have a balanced diet are getting all the essential food groups in the correct proportions.

These food groups are as follows:

1 **Carbohydrates:** These include simple sugars such as **glucose** and **lactose** (the sugar in milk) and the more complex carbohydrates (**cellulose**, **glycogen** and **starch**).

Starch is a very important part of the human diet and it is broken down to glucose during digestion. In our bodies we store carbohydrate as glycogen and this can be broken down to glucose when our sugar reserves are low. However, humans are unable to digest cellulose (see the later section on fibre for the role of cellulose in the diet).

Carbohydrates provide **energy** – sugar is a fast-acting energy source and starch is a slow-release source. Examples of foods rich in sugar include biscuits, cakes, jam and fizzy drinks. Potatoes, rice, pasta and bread are all rich in starch.

Figure 3.4a Foods rich in carbohydrate

Figure 3.4b Foods rich in protein

Figure 3.4c Foods rich in fat

Questions

4 Suggest why a few centuries ago many sailors taking part in long voyages suffered from bleeding gums.
5 Suggest why people living in tropical regions are less likely to suffer from vitamin D deficiency than those living in polar regions.

2 **Protein:** Protein provides the building blocks for the **growth and repair** of cells but can be used for energy when reserves of carbohydrate and fat are low. When protein is digested it is broken down to **amino acids**. Good examples of protein-rich foods are lean meat, fish and egg white.

3 **Fat:** This is an excellent **energy store**, providing double the energy per gram of carbohydrate and protein. Fat is also important in providing **insulation** and **protects** some organs. When fat is digested it is broken down into **fatty acids** and **glycerol**. Fat-rich foods include streaky bacon, cheese and lard.

Carbohydrates, proteins and fats all contain the elements **carbon, hydrogen** and **oxygen** but all proteins also contain the element **nitrogen**.

4 **Vitamins** and **minerals** are needed for the healthy functioning of our bodies. We only need relatively small amounts of each of these in our diet but if they are missing the symptoms (signs) of deficiency (shortage) become apparent. Thankfully most people in Western societies have access to enough of each of the essential vitamins and minerals. However, many people in underdeveloped parts of the world, such as many areas of Africa, suffer from deficiencies in vitamins or minerals, and some simply do not have enough food of any type.

Some of the essential **vitamins** with their sources and roles in the body are included in the table below.

Vitamin	Good sources	Role in the body	Problems if we don't absorb enough (deficiency symptoms)
C	Citrus fruits (oranges, lemons, limes, etc.) and vegetables such as broccoli and peppers	Keeps the cells of the body in good working order, particularly teeth and gums. Keeps blood vessels strong	Scurvy – bleeding gums
D	Fish liver oil, liver, eggs and milk (can also be made by the skin when sunlight shines on it)	Essential for normal growth – helps in the development of bones and teeth	Rickets – poor bone devlopment

Minerals such as calcium and iron are also essential for healthy living. The sources, functions and deficiency symptoms of these minerals are included in the table below.

Mineral	Good sources	Role in the body	Problems if we don't absorb enough (deficiency symptoms)
Calcium	Milk, cheese, cereals, vegetables	Essential for bone and teeth development, nerve and muscle action and blood clotting	Poor development of bones and teeth
Iron	Liver, red meat, spinach	Needed for making haemoglobin in red blood cells. The haemoglobin transports oxygen around the body	Anaemia and low energy levels caused by the body not getting enough oxygen

Figure 3.5 Foods rich in fibre

5 **Fibre:** Essential for the efficient working of the digestive system and for preventing constipation. The cellulose from plant cell walls is the main constituent of fibre. Foods rich in fibre include wholemeal bread and green vegetables.

6 **Water:** Important as a transport medium and a component of cytoplasm and body fluids and for its use in chemical reactions.

Exam questions

1 a) The energy released from foods can be compared using the apparatus shown in the diagram.

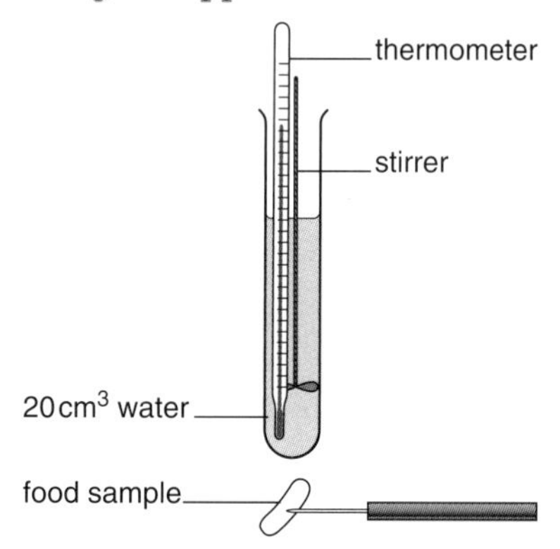

i) Using the apparatus and any additional materials, describe how you would collect data to calculate and compare the energy released from equal masses of biscuit and bacon.
In this question you will be assessed on your written communication skills including the use of specialist science terms. *(6 marks)*

ii) How would you expect the results for the biscuit and bacon to differ? *(1 mark)*

2 a) The graph shows the daily energy requirements for humans at different ages.

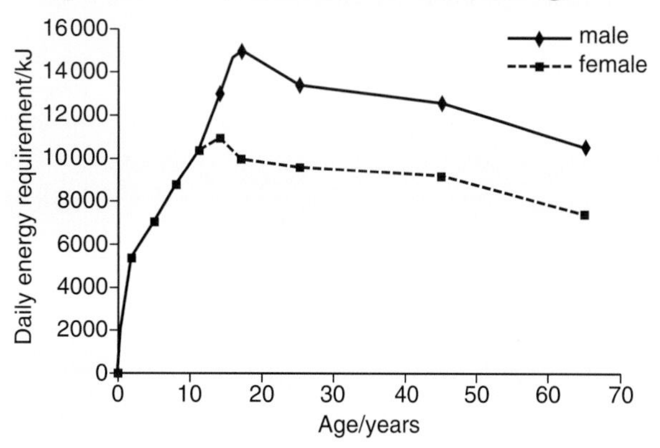

i) Describe the trend for males aged:
0–17 years
20–65 years *(2 marks)*

ii) Why does the body need energy? *(1 mark)*

iii) Why are the female energy requirements lower than those for males? *(1 mark)*

iv) Give **one** other factor which affects the energy requirements of adults. *(1 mark)*

v) Describe, with reference to the graph, what a starvation diet would be for a 25-year-old female. *(2 marks)*

b) The apparatus shown was used to calculate the energy in a piece of pasta.

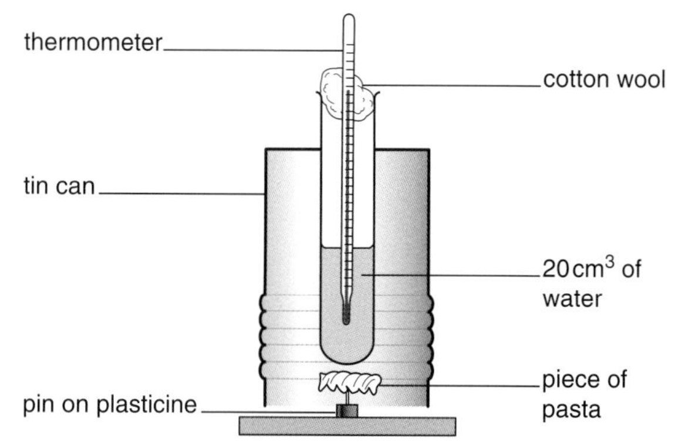

4.2 J of energy will increase the temperature of 1 cm^3 of water by 1 °C. The temperature of the water increased by 25 °C during the experiment.

i) Use this information to help calculate the energy in the piece of pasta. Show your working. *(2 marks)*

ii) Suggest **one** reason why the results of this experiment may be considered unreliable. *(1 mark)*

The table summarises the nutritional information for pasta.

	Per 100 g	Percentage of daily requirement
Energy	1515 kJ	19.1
Protein	12.3 g	
Carbohydrate Starch	74.0 g 69.6 g	25
Fat Cholesterol	1.0 g 0.0 g	2.0 0.0
Fibre	3.0 g	12
Salt	Less than 0.1 g	1.7
Calcium Iron	18.0 mg 1.0 mg	2.0 6.0

c) Using **two** pieces of evidence from the table, explain why eating pasta may be an advantage as part of a balanced diet. *(4 marks)*

3 a) Give **two** ways food is used in the body. *(2 marks)*

b) The diagram shows two food tests.

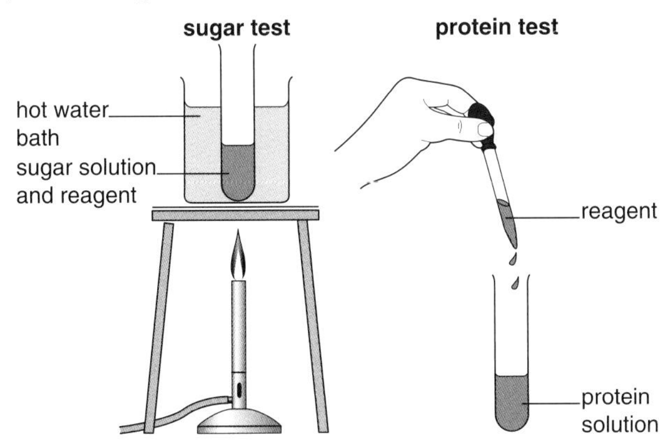

i) Use the diagram to help describe the test for sugar. *(2 marks)*

ii) Describe the colour when sugar is present. *(1 mark)*

iii) Name the reagent used to test for protein and give the colour change when protein is present. *(3 marks)*

4 The diagram shows the apparatus used to test fruit juice for vitamin C content.

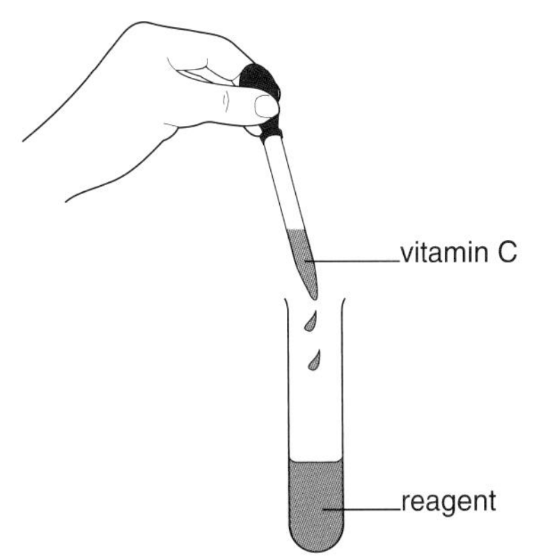

a) Name the reagent used to test for vitamin C. *(1 mark)*

b) Describe the colour change if vitamin C is present. *(1 mark)*

The table shows the results for three different juices.

Type of juice	Number of drops of fruit juice required to change colour of reagent
Blackcurrant	3
Orange	5
Boiled orange	18

c) Which juice contained the most vitamin C? *(1 mark)*

d) Suggest **two** reasons why boiling fruit and vegetables reduces their vitamin C content. *(2 marks)*

5 Carbohydrates contain the element carbon.

a) i) Name the **other two** elements found in carbohydrates. *(2 marks)*

ii) Name the carbohydrate:
found in fibre
used in respiration. *(2 marks)*

b) A 50 g serving of pasta contains 756 kJ of energy.

i) Calculate the energy content of 100 g of pasta. *(1 mark)*

ii) Suggest which of the foods below would provide most energy.
potatoes butter cabbage *(1 mark)*

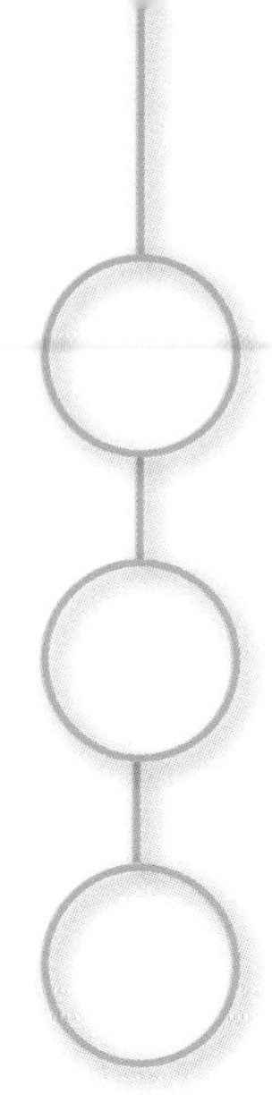

Digestion and enzymes

Most of the food we eat is in the form of large, complex, insoluble molecules. We need to digest and break down these molecules so that they are soluble and small enough to be absorbed into the blood system where they can be used by the body.

Digestion can be defined as the breakdown of large, complex, insoluble molecules into small, simple, soluble ones.

The alimentary canal (gut) is the organ system in the body that is adapted to enable the body to carry out digestion (and absorption) so that we can gain all the nutrients we need from our food.

The alimentary canal

Figure 4.1 shows the human alimentary canal and its associated organs.

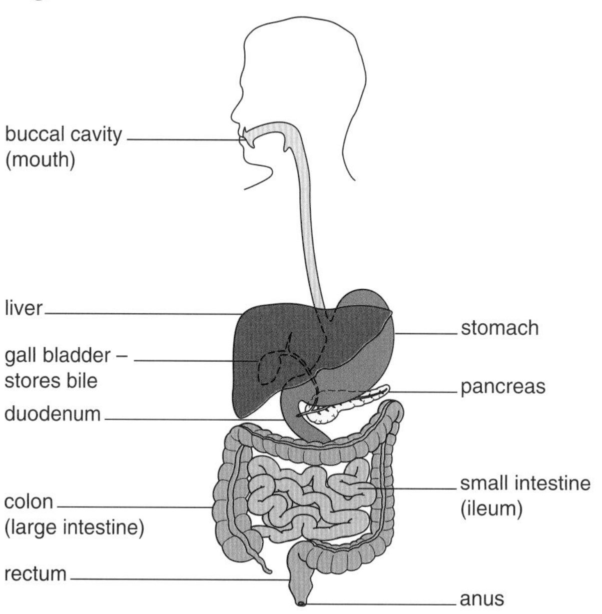

Figure 4.1 The alimentary canal

1 – The buccal cavity

Food enters the **mouth (buccal cavity)** and **mechanical digestion** occurs through the action of the teeth. In the mouth the **chemical digestion** of food also takes place through enzyme action.

Note: As humans are neither true carnivores nor herbivores our teeth are adapted for a wide range of foods including both plant and animal products.

Amylase produced in the salivary glands breaks down **starch** into **sugar**.

2 – The stomach

When the food reaches the stomach, the stomach produces **gastric juice**. The gastric juice contains **protease** enzymes that break down **proteins** into **amino acids**. It also contains **acid**, which provides an acidic pH in the stomach (necessary for the action of stomach protease) and also destroys microorganisms taken in with the food.

3 – The liver

The liver produces **bile**, which is then stored in the **gall bladder**. When food enters the top part of the small intestine (the duodenum) the bile is released and **neutralises** the acidic food passing through from the stomach. It makes the upper part of the small intestine slightly alkaline. Bile also **emulsifies fats**. Emulsification is the breakdown of large fat globules into smaller fat globules, which increases the surface area of the fat to make it easier for the lipase enzymes to break it down.

4 – The small intestine (duodenum)

The duodenum is the section of gut immediately below the stomach and is the main site of enzyme action.

Pancreatic juice produced by the pancreas contains additional amylase and protease to complete the starch and protein digestion that began in the mouth and stomach. Additionally, **lipase** for the breakdown of **fats** into **fatty acids** and **glycerol**, and **carbohydrases** for the breakdown of a range of **carbohydrates** into their constituent sugars are also produced in the pancreatic juice.

The walls of the duodenum itself also produce **intestinal juice** containing amylase, protease, lipase and carbohydrases.

5 – The ileum (small intestine)

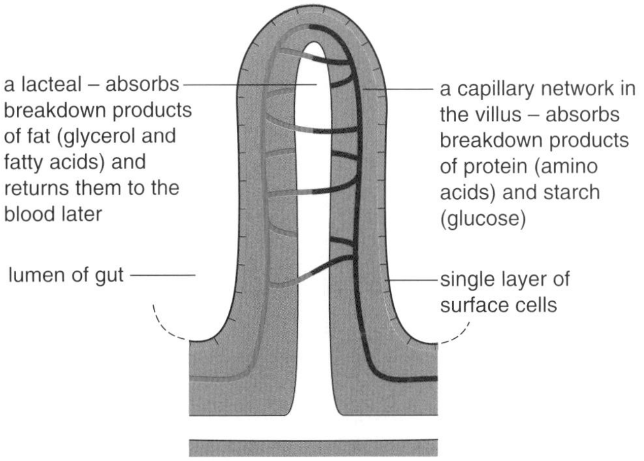

The main site for the **absorption** of the digested food products is the ileum (although some digestion takes place here too, e.g. proteases, lipases and amylases break down protein, fats and starch respectively). It is adapted for the process of absorption in many ways, including its great **length** (over 6 metres) and the presence of many **folds** (or twists) along its length that greatly increase the **surface area** for the absorption of food. The process is aided by the **good blood supply** and the **thin and permeable membranes**.

The presence of microscopic outgrowths called **villi** on the inner surface of the ileum also increases the surface area. Figure 4.2 shows a section through a villus.

Figure 4.2 Structure and function of a villus

The diagram shows that the millions of villi that line the inner surface (the part in contact with the food) aid absorption due to the presence of:

* the **excellent blood supply** – each villus has an extensive capillary network, which means that the whole small intestine is well supplied with blood to transport the absorbed products of digestion.
  Note: the lacteal is part of a system of tubes that absorb the breakdown products of fat before returning them to the blood.
* the **thin** and **permeable surface** lining each villus – this is due to the small number of cells between the lumen of the gut and the capillaries and lacteals.

The villus is further adapted for the efficient absorption of digested food through their finger like shape which increases **surface area** and the single layer of **surface (epithelial) cells**.

6 – The colon (large intestine)

In this part of the alimentary canal, **water** is absorbed from the gut into the blood. The colon is adapted by having a **large surface area**.

Question

1 Describe the processes of protein digestion and absorption in the alimentary canal.

7 – The rectum and anus

The regions where the **storage** and **removal** of faeces take place respectively.

The action of enzymes

Enzymes are biological catalysts that speed up the rate of biological reactions. Enzymes are very important in digestion and the substrates (molecules broken down) and products of the main enzymes are described in earlier sections. All enzymes are proteins.

Question

2 Copy and complete the following table.

Enzyme	Substrate	Product	Region(s) in gut where enzymes act
Amylase			
Protease			
Lipase			
Carbohydrase			

Enzymes work by the substrate fitting snugly into the **active site** of the enzyme. This tight fit then enables the enzyme to catalyse the reaction and split the substrate into its products – the role of the enzyme is to speed up the rate of reactions.

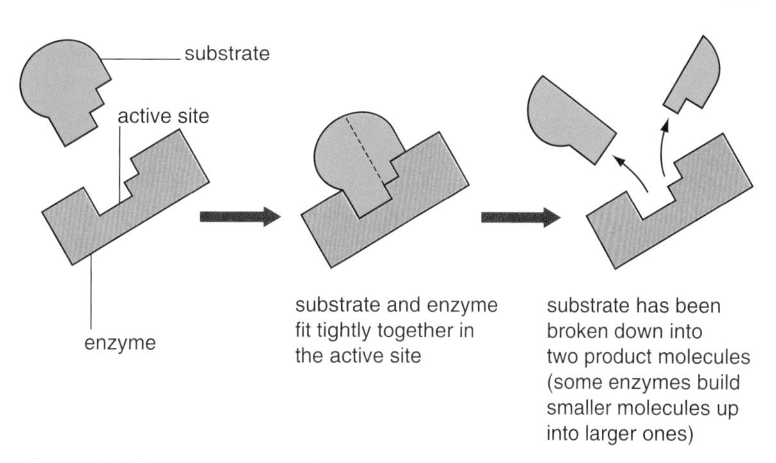

Figure 4.3 How enzymes work

The action of enzymes as described in Figure 4.3 is referred to as the **'lock and key model'** due to the importance of the tight fit between the enzyme's active site and the substrate. This 'lock and key model' can explain the principle of **enzyme specificity** – each enzyme is specific in that it will only work on its normal substrate. For example, only starch will fit into the active site of amylase – other molecules such as proteins cannot.

The effects of temperature and pH on the action of amylase

Temperature and pH affect the activity of all enzymes. If enzymes do not have their optimum temperature or pH they work less effectively. Figures 4.4 and 4.5 show how changing the temperature and pH affect the activity of amylase.

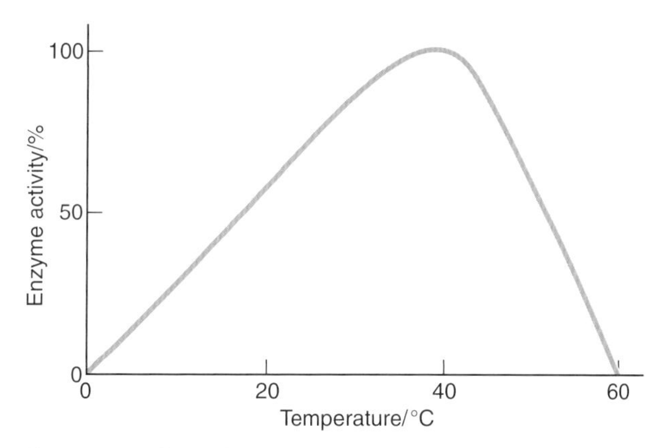

Figure 4.4 Effect of temperature on enzyme activity

At low temperatures the reduced kinetic energy of the enzymes and substrates lead to reduced rates of collision. The maximum rate is the optimum rate and at increasingly higher temperatures (or away from the pH optimum) denaturation increasingly occurs and is due to the irreversible change to the shape of the enzyme active site.

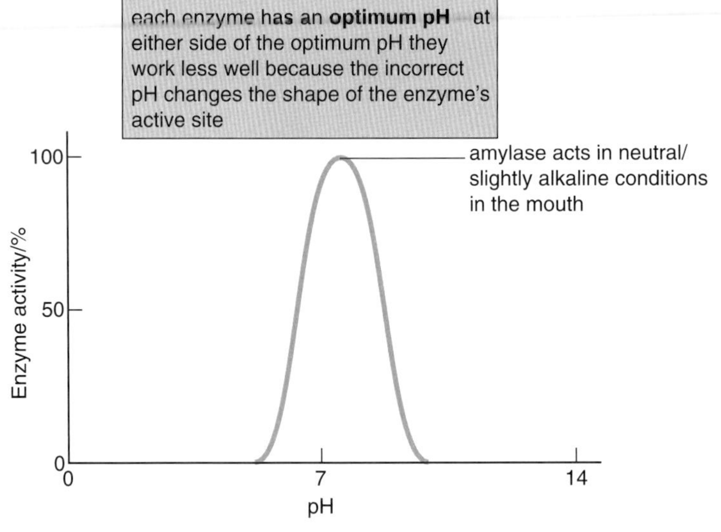

Figure 4.5 Effect of pH on enzyme activity

The effect of enzyme concentration on enzyme activity

The more enzymes there are, the faster the enzyme reaction. This is because there are more active sites for substrates to attach to. This applies up to a limit, when the rate levels off because there are not enough substrate molecules to react with the extra enzymes.

Question

3 Use the graph opposite to describe and explain the results at X and Y.

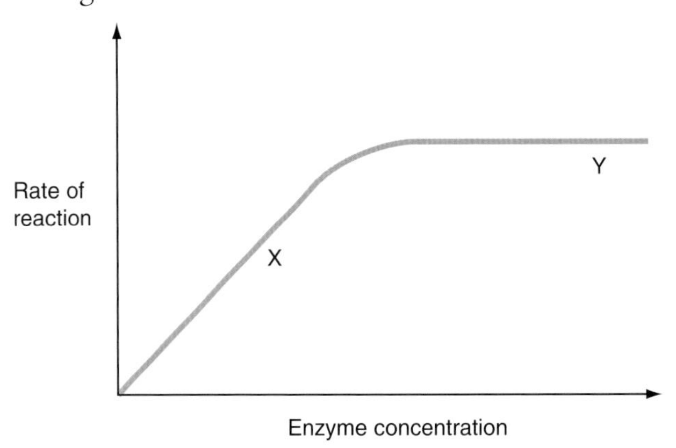

Figure 4.6 Effect of enzyme concentration on enzyme activity

Commercial enzymes

Search ▶ washing powders and enzymes

Many enzymes have commercial uses. For example, many 'biological' washing powders have enzymes for breaking down difficult-to-remove stains. These enzymes are **thermostable** – they can work at a wide range of temperatures – and they break the complex, large and insoluble stains down into small, soluble molecules that dissolve in the water.

Questions

4 Blood (containing protein) and grass (containing cellulose and other complex carbohydrates) can cause stains on clothes that are difficult to remove.
   a) Name the type of enzymes that can be used to remove these stains.
   b) Suggest two advantages of the enzymes in many washing powders being able to work effectively at less than 40 °C.

Exam questions

1 The diagram shows the digestive system and associated organs.

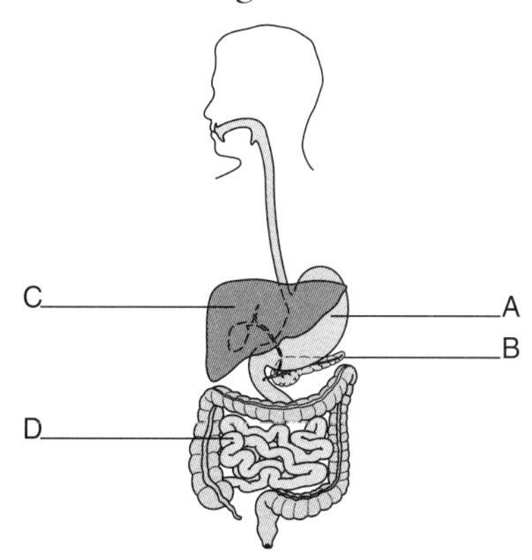

a) Explain why it is important that food is digested. *(1 mark)*

b) In which of the labelled parts on the diagram (A, B, C or D) does protein digestion begin? *(1 mark)*

c) Copy and complete the following table to show the types of food broken down by the enzymes named in the table and the products of digestion. *(2 marks)*

Enzyme	Food broken down	Product
Protease		
Amylase		

2 Explain how the structure of the ileum is adapted to the function of absorption of digested food molecules.
In this question you will be assessed on your written communication skills including the use of specialist science terms. *(6 marks)*

3 The activity of the enzyme lipase was investigated at three different temperatures. The initial pH of the reaction mixture was 7.0 in all cases. The final pH was measured after 60 minutes and the results are presented in the table.

Temperature/°C	pH of reaction mixture after 60 minutes
20	6.8
35	5.8
50	7.0

a) Suggest what product produced by the action of lipase affects pH. *(1 mark)*

b) Describe and explain the effect of temperature on lipase activity. *(3 marks)*

4 a) i) Define an enzyme. *(1 mark)*

ii) Copy the diagram and label the active site of the enzyme. *(1 mark)*

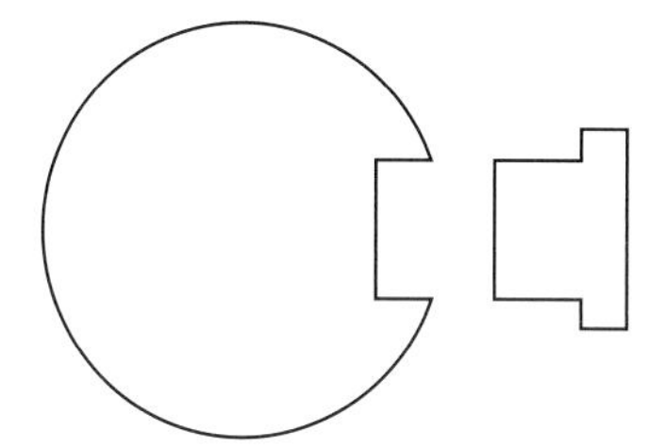

iii) What is the term given to this model of enzyme action? *(1 mark)*

b) David wants to test the effect of amylase on starch so he sets up an experiment as shown below.

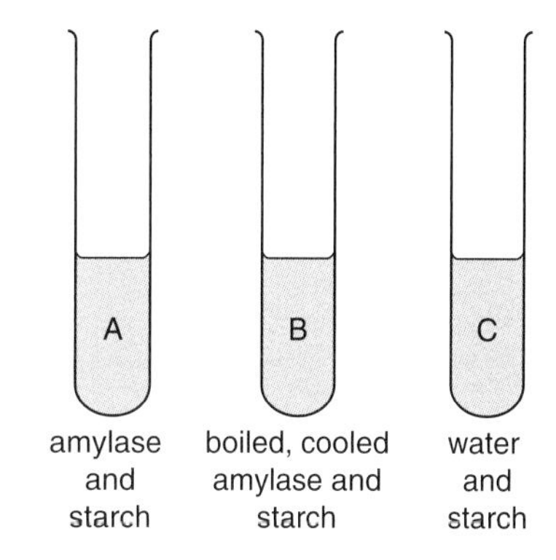

David placed three test tubes and solutions separately in a water bath at 30 °C for 5 minutes. He then set up the three test tubes as shown in the diagram. He took samples from the test tubes at different times and tested them with iodine.
Describe and account for David's results for test tubes A, B and C.
In this question you will be assessed on your written communication skills including the use of specialist science terms. *(6 marks)*

5 Enzymes are present in the human digestive system and are used commercially in biological washing powder.

a) What is an enzyme? *(2 marks)*

Proteases are one type of enzyme used in biological washing powder. An investigation was carried out to measure the performance of two different types of protease, protease X and protease Y, over a range of temperatures.

The results are shown in the following table.

Temperature/°C	Mass of protein broken down/ arbitrary units	
	Protease X	Protease Y
0	0	0
5	10	1
10	30	3
15	18	8
20	2	19
25	0	22
30	0	29
35	0	31
40	0	30
45	0	29
50	0	23
55	0	0
60	0	0

b) Use the data from the above table to draw a graph, comparing the effect of temperature on the two protease enzymes. *(3 marks)*

c) One possible conclusion of this investigation is that the protease enzymes would be equally good at stain removal.

i) What evidence from the graph supports this conclusion? *(1 mark)*

ii) What evidence does not support this conclusion? *(1 mark)*

d) i) Based on this investigation, explain which of the protease enzymes would produce a more environmentally friendly washing powder. *(2 marks)*

ii) The manufacturers used a combination of protease X and protease Y in their washing powder. Why would they have decided on this? *(1 mark)*

iii) Explain why a washing powder containing only protease enzymes would not remove grease and fat stains. *(2 marks)*

The diagram shows a model of part of a protein molecule and a molecule of protease X.

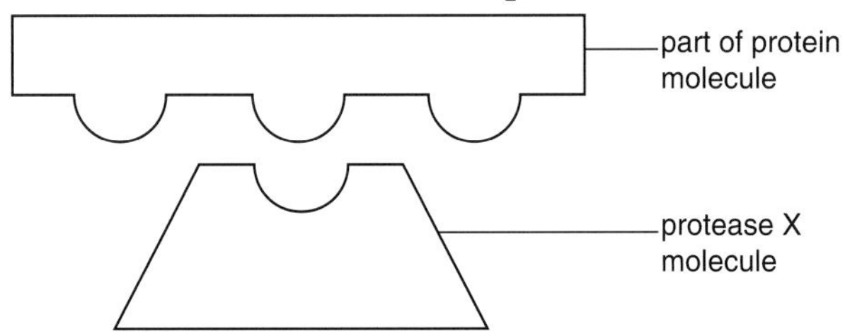

e) Use information from the graph and the model above to describe and explain the activity of protease X.
In this question you will be assessed on your written communication skills including the use of specialist science terms. *(6 marks)*

5 Breathing and the Respiratory System

The function of the respiratory system is to provide the tissues with oxygen and remove carbon dioxide to facilitate the process of respiration.

Figure 5.1 shows the main parts of the respiratory system.

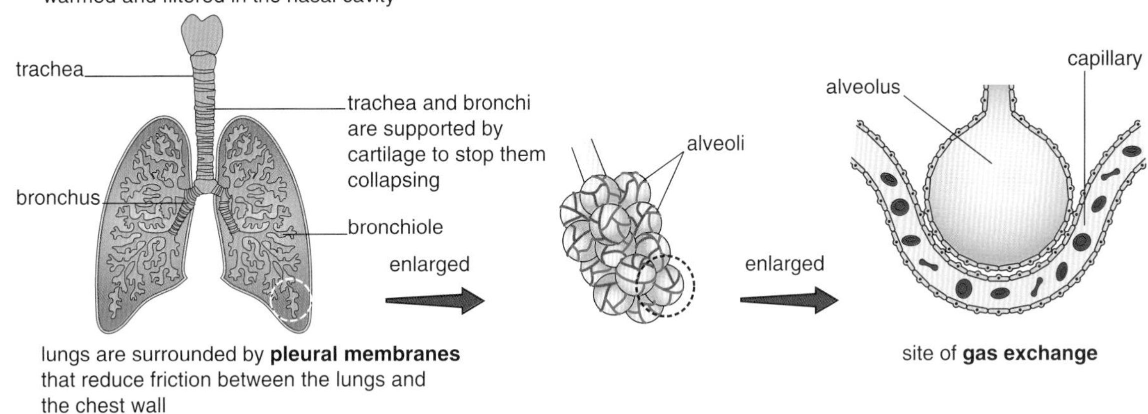

Figure 5.1 The respiratory system

Note how air can pass into the lungs through the sequence:

nasal cavity → trachea → bronchus → bronchiole → alveolus

▶ The process of breathing

The process of breathing can be examined using a lung model. Many lung models resemble the bell jar apparatus shown in Figure 5.2. Lungs are represented by balloons and the diaphragm is represented by a thin sheet of rubber.

Figure 5.2 The bell jar lung model

When the rubber representing the diaphragm is pulled down the lungs inflate and they deflate as the rubber sheet is released.

This model demonstrates the following key features of the breathing process:

* As the diaphragm (rubber sheet) moves down, the **volume** inside the glass jar **increases**.
* This causes the **pressure** inside the glass jar to **decrease**.
* This causes **air to enter** the lungs until the pressures inside and outside the bell jar become equal.

In reality the process of breathing is more complicated than is shown by the model.

The diaphragm and intercostal muscles, together with the other parts of the respiratory system shown in Figure 5.1, are involved in the processes of inhaling and exhaling (breathing). Breathing ensures that fresh air rich in oxygen is brought into the lungs and carbon dioxide-rich air is expelled from the lungs. Figure 5.3 shows the process of breathing in humans.

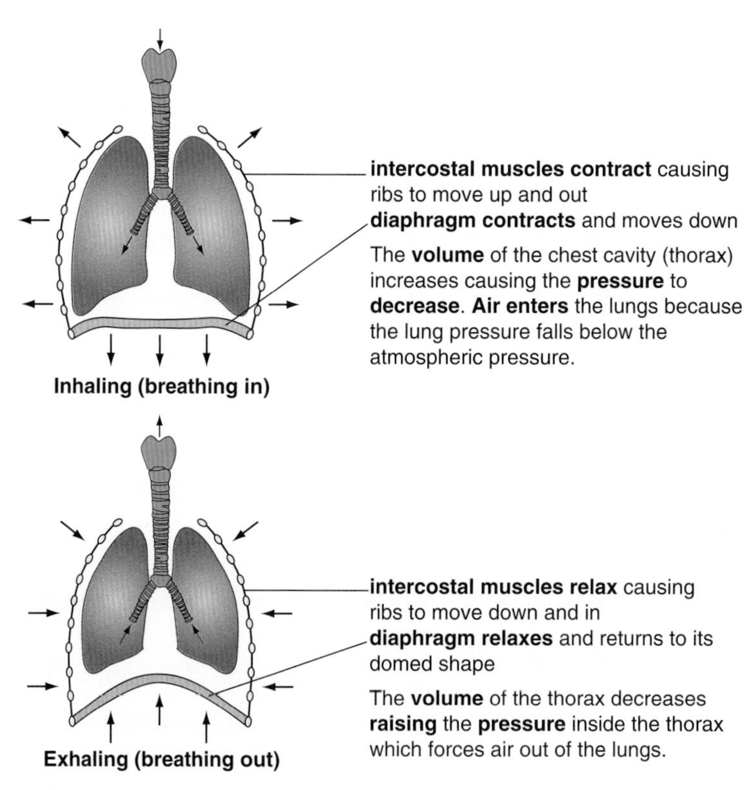

Figure 5.3 Inhaling and exhaling

There are a number of important differences between the lung model in Figure 5.2 and the process of breathing in humans. These include:

* In humans the ribs move out and in and work together with the diaphragm in changing the volume of the thorax. In the lung model only the diaphragm (rubber sheet) is involved.

Question

1 Suggest what would happen in the model if there was a crack in the glass allowing air to enter and leave. Explain your answer.

* In humans the diaphragm is normally a domed shape – it flattens when breathing in (in the bell jar it starts as a flat shape and is pulled down).
* The space between the lungs (balloons) and the chest wall (glass jar) is much greater in the model than between the lungs and ribcage in reality.

Pleural membranes line the inside of the chest wall (ribs) and the outside of the lungs. Their role is to reduce friction during breathing.

The space between the membranes (the pleural cavity) usually contains a small amount of **pleural fluid**, which further helps to reduce friction during breathing.

Inhaled and exhaled air – what is the difference?

The following table summarises the differences between inhaled and exhaled air.

Gas	Inhaled air/%	Exhaled air/%	Explanation of change
Oxygen	21	16	Some of the oxygen in inhaled air has diffused from the alveoli into the blood and is transported round the body to be used by the body's cells for respiration.
Carbon dioxide	0.04	4	Carbon dioxide produced in respiration by the body cells is transported back to the lungs, diffuses from the blood into the alveoli and is expelled in exhaled air.
Nitrogen	78	78	Nitrogen is not used in respiration.

▶ Respiratory surfaces

The respiratory surfaces are the parts of the body where **gas exchange** takes place between the atmosphere and the blood. In humans gas exchange takes place in the alveoli. Oxygen diffuses into the blood and carbon dioxide diffuses from the blood into the alveoli, where it is then breathed out.

Respiratory surfaces are adapted in a number of ways. For example, in humans they have:

* **a large surface area** – there are many alveoli in each lung and each alveolus has a large surface area. Together this gives a gas exchange surface (where the alveolar walls are in contact with blood capillaries) in humans of many square metres
* **thin walls with short diffusion distances** – Figure 5.1 shows that there are only two layers of cells separating the oxygen in the alveolus from the red blood cells. This means that there is a short diffusion distance for the gases involved
* **moist walls** – these help the gases to pass through the respiratory surfaces because the gases dissolve in the moisture
* **permeable surfaces** – the moist, thin walls make the respiratory surfaces permeable

* **a good blood supply** – alveoli are surrounded by capillaries to ensure that any oxygen diffusing through is carried around the body. This also ensures that carbon dioxide is continually taken back to the lungs
* **a diffusion gradient** – the process of breathing ensures that there is a large diffusion gradient that encourages oxygen to diffuse into the blood and carbon dioxide to diffuse from the blood into the alveoli. When fresh air rich in oxygen is breathed in, it makes the concentration of oxygen in the alveoli higher than that in the capillary and therefore oxygen diffuses from the alveoli into the capillaries.

Questions

2 Draw a diagram of an alveolus and a surrounding blood capillary.

3 a) On a diagram of an alveolus draw an arrow to show the movement of oxygen.
b) Using only information in your drawing, describe how the alveolus is adapted for gas exchange.

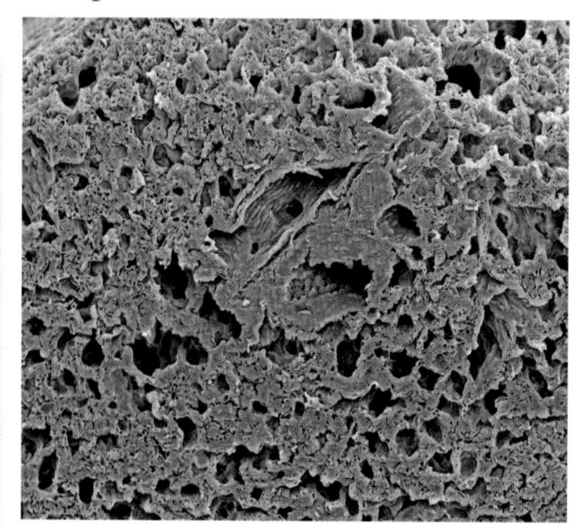

Figure 5.4 Section through lungs showing the presence of many air spaces

The same principles apply to the respiratory surfaces in plants. The main respiratory surfaces in plants are the cells surrounding the air spaces in the leaves. Because there are a lot of cells in contact with the air spaces, there is a large surface area and the cell membranes (where gas exchange takes place) are thin, moist and permeable.

Investigating the effect of exercise on breathing rate

To investigate the effect of exercise on breathing rate and recovery rate you first need to calculate your breathing rate at rest. Then carry out vigorous exercise for a short period of time. Measure your breathing rate immediately when you stop and then at intervals, for example every minute, until it returns to normal. The time taken for the breathing rate to return to normal can be referred to as the recovery time. Compare your results with other members of your class.

You should be able to use the results you obtain to discuss the following points.

* Is there a link between levels of fitness and breathing rate?
* Is there a link between levels of fitness and recovery time?
* If a pupil became fitter over time, how would this affect his or her recovery time?

Question

4 a) Describe how the breathing rate changes during exercise and explain why it is important that it does.
b) John is going on a 20-minute run. Sketch a graph to show changes in his breathing rate from 10 minutes before he starts running until 10 minutes after he has finished running.

The word equation for aerobic respiration is:

glucose + oxygen → carbon dioxide + water + energy

The balanced chemical equation is:

$$C_6H_{12}O_6 + 6O_2 \rightarrow 6CO_2 + 6H_2O + \text{energy}$$

The energy released in respiration can be used to produce heat in the body, movement, growth, reproduction and **active transport**.

Aerobic and anaerobic respiration

Respiration using oxygen is **aerobic** and respiration without oxygen is **anaerobic**. Most living organisms must respire using oxygen (aerobically) but many can respire anaerobically in some circumstances.

The word equations for anaerobic respiration highlight the differences between aerobic and anaerobic respiration.

Anaerobic respiration in yeast:

glucose → carbon dioxide + alcohol + a small amount of energy

Similarly, human muscles can respire anaerobically for a short period of time if aerobic respiration does not meet their energy needs. This is likely to happen during strenuous exercise.

Question

5 Use the information given to suggest why humans can only respire anaerobically for a short period of time.

Anaerobic respiration in muscle:

glucose → lactate + a small amount of energy

Demonstrating anaerobic respiration in yeast

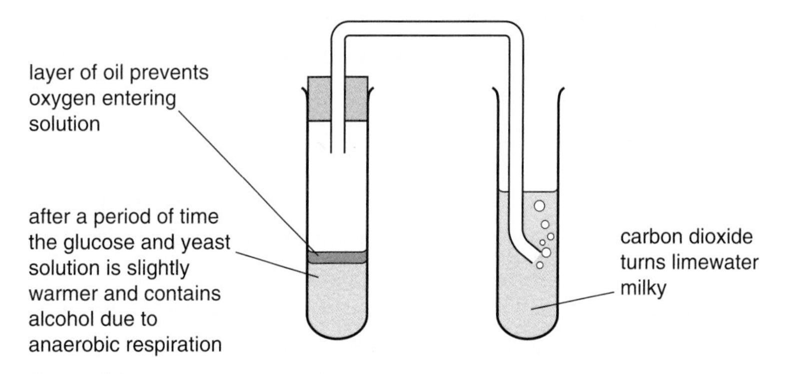

Figure 5.5 Demonstrating anaerobic respiration in yeast

Question

6 Using yeast as an example, describe the differences between aerobic and anaerobic respiration.

Exam questions

1 The diagram shows a lung model.

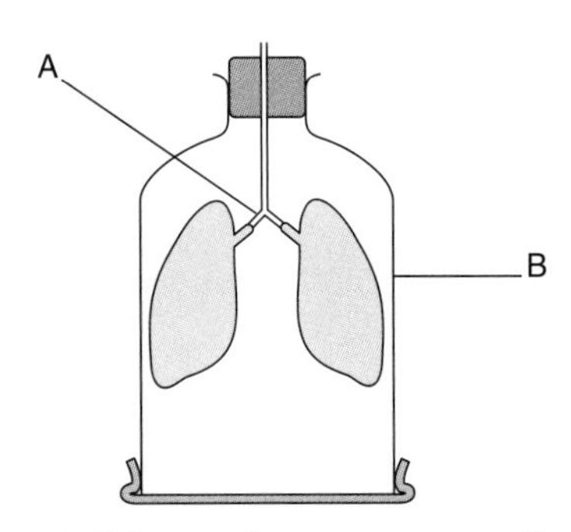

a) Name the structures of a human lung modelled by parts A and B. *(2 marks)*

b) Copy and complete the flowchart, showing the pathway of a molecule of oxygen through the lung and into the blood. *(3 marks)*

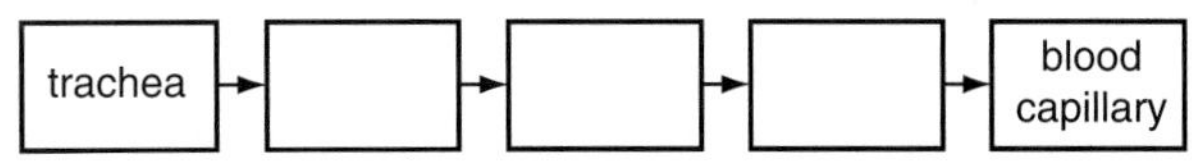

c) Describe the part played by the contraction of the intercostal muscles in the breathing mechanism. *(3 marks)*

2 a) Describe the mechanism of breathing. *(3 marks)*

The graph shows changes to the volume of air breathed and the number of breaths taken while walking.

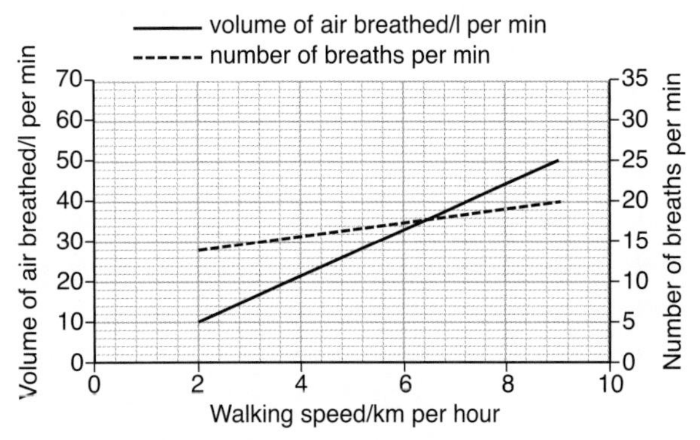

b) Using values from the graph, calculate the volume of each breath when walking at 2 km per hour. *(1 mark)*

c) The volume of air taken in each breath changes as walking speed increases. Explain how this helps gas exchange in the lungs. *(2 marks)*

3 a) Copy and complete the table to compare aerobic and anaerobic respiration of yeast. Write **Yes** or **No** in each of the empty boxes. *(4 marks)*

	Respiration in yeast	
	Aerobic respiration	Anaerobic respiration
Uses oxygen		
Releases energy		
Produces ethanol		
Produces carbon dioxide		

b) Where does aerobic respiration take place in the human body?
Describe what happens to the substances produced. *(2 marks)*

4 The diagram shows the apparatus used to show the effect of temperature on anaerobic respiration in yeast.

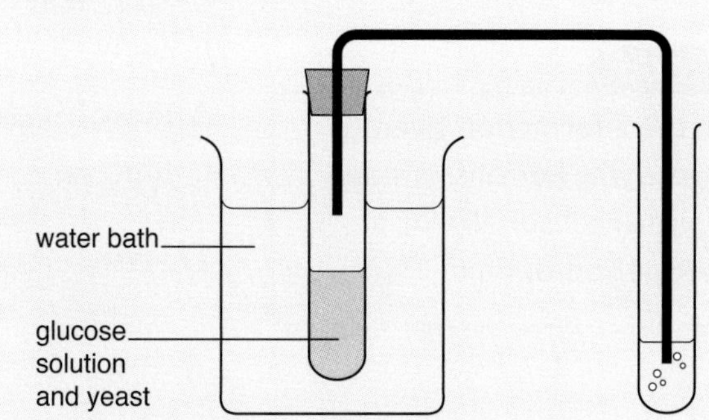

a) i) When setting up the experiment, suggest how you would ensure that the glucose solution is anaerobic at the start of the experiment. *(1 mark)*

ii) Describe how you would ensure that the glucose solution remains anaerobic throughout the experiment. *(1 mark)*

b) Describe how you would use the apparatus to show the effect of temperature on the rate of anaerobic respiration. *(2 marks)*

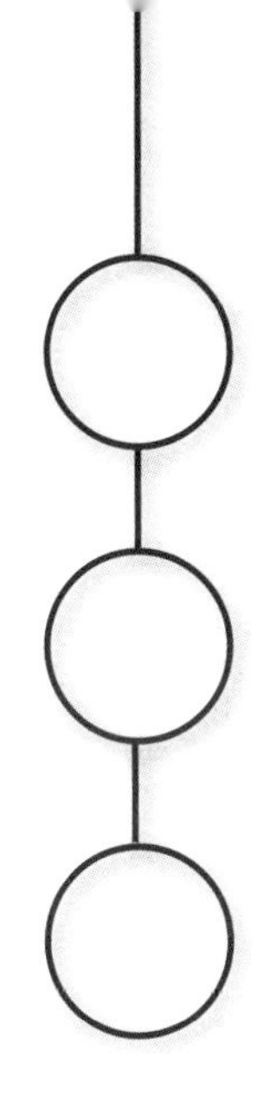

6 The Nervous System and Hormones

Complex living organisms (particularly animals) are aware of their surroundings and can respond appropriately to change. Animals can also respond to changes in their bodies (internal environment). This awareness and ability to respond is due to the presence of a **nervous system** and also **hormones**.

▶ Responding to the environment – the nervous system

We are able to respond to the environment and anything that we respond to is called a **stimulus**. In animals each type of stimulus affects a **receptor** in the body. There are many types of receptors, each responding to a particular type of stimulus or sense (e.g. sight, sound, touch, taste and smell). If a receptor is stimulated it may cause **effectors** (e.g. muscles) to produce a **response**.

stimulus → **receptor** → **effector** → **response**

The flowchart above is a simplification because it suggests that we will automatically produce a response when we are stimulated. For example, if we hear a sound (the stimulus) we might respond or not, depending on what the sound is.

Co-ordination

In reality the receptors and effectors are linked by a **co-ordinator**. This co-ordinator is usually the **brain** but may also be the **spinal cord**. Together the brain and spinal cord are known as the **central nervous system** (**CNS**).

Nerve cells or **neurones** link the receptors and effectors to the co-ordinator. A neurone carries information as small electrical charges called **nerve impulses**. The brain acts as a filter and determines which receptors link up with which effectors – and even whether or not a particular stimulus brings about a response.

A more complete flowchart is:

stimulus →	**receptor** →	**brain** →	**effector (muscle)** →	**response**
Jane texts John	*John's eye reads text*	*John thinks what to do*	*John types a reply*	*John texts Jane*

The overall total of our responses to the environment around us is described as our **behaviour**.

Sometimes our receptors are grouped together into complex sense organs. Examples include the nose (smell), the ear (sound) and the eye (sight).

▶ The eye

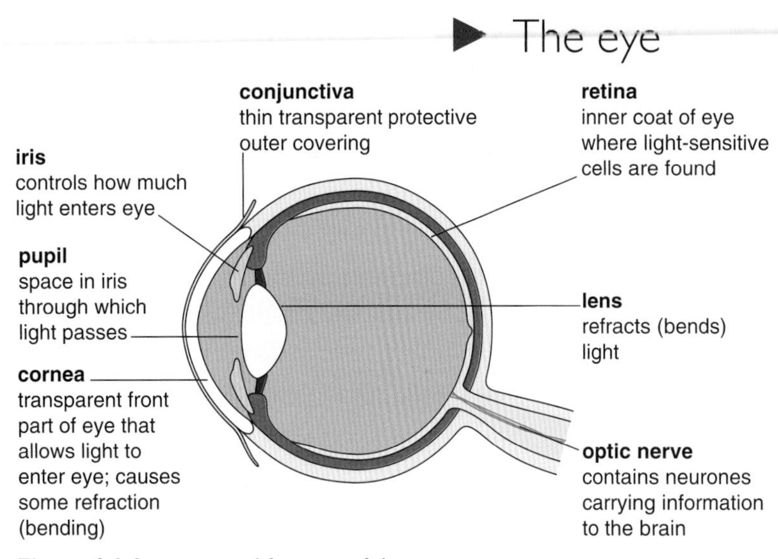

Figure 6.1 Structure and function of the eye

The eye is a specialised sense organ that contains receptors that are sensitive to light. Figure 6.1 shows the main parts of the eye and their functions.

The watery fluid between the cornea and the lens (**aqueous humour**) and between the lens and the retina (**vitreous humour**) keeps the lens in shape and allows light through.

Focusing the image

As light rays enter and pass through the cornea, some bending (refraction) of light takes place. Further bending takes place as the light passes through the lens. By adjusting the thickness of the lens, light rays can be focused on the retina, irrespective of their angle as they enter the eye. Figure 6.2 shows the role of the lens in focusing light rays from distant and near objects on the retina.

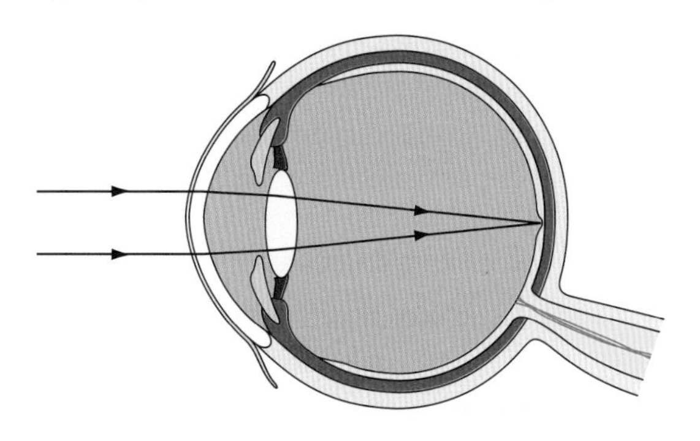

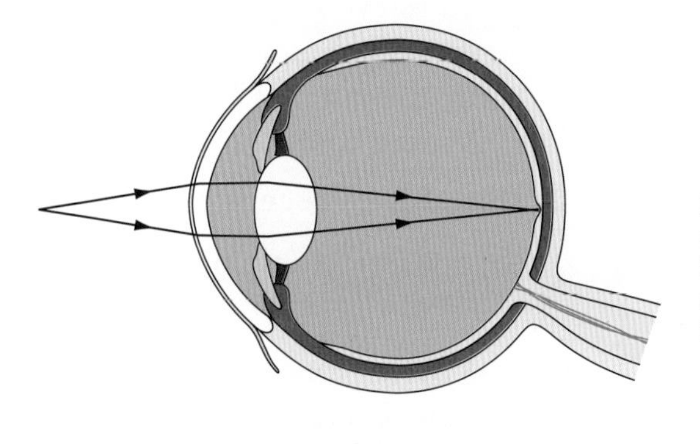

Figure 6.2 How an image is focused on the retina

How does the lens change shape? (accommodation)

The **ciliary muscle** is a ring of muscle that surrounds the lens. The lens is attached to the ciliary muscle by **suspensory ligaments** that resemble small pieces of thread. If the ciliary muscle relaxes it springs out to give a bigger diameter. When this happens the suspensory ligaments pull the lens and it becomes thinner. The opposite happens to make the lens fatter. The ciliary muscle contracts to form a tighter circle with a smaller diameter. The suspensory ligaments then relax and with less pressure on the lens it is able to spring back to its original thicker shape.

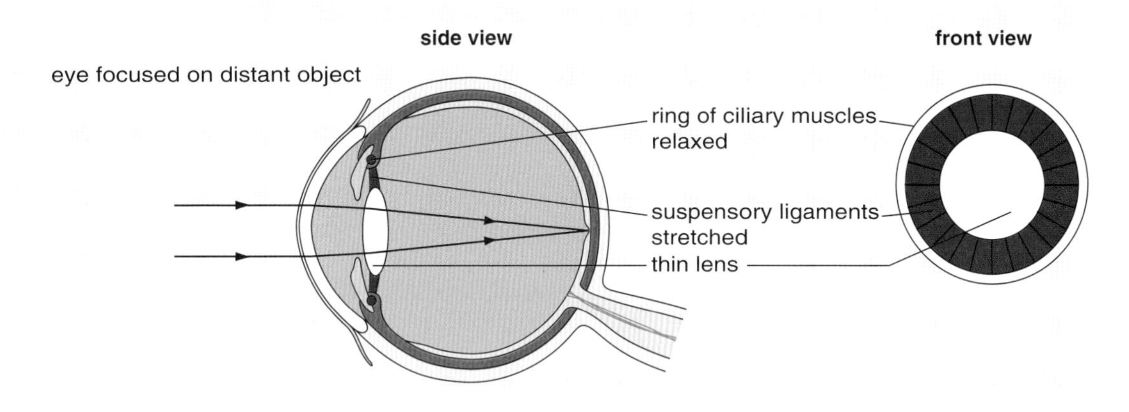

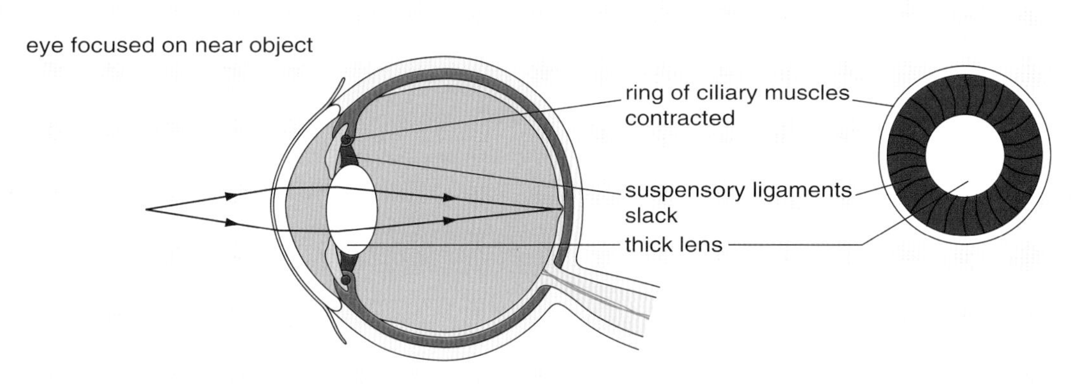

Figure 6.3 Changing the shape of the lens

Question

1 Describe and explain how the eye is able to focus on a close-up object.

Controlling the amount of light that enters the eye

It is important that the correct intensity of light enters the eye and reaches the retina. Too little or too much light will prevent an image being formed. In addition, too much light can damage the sensitive light receptor cells in the retina. Dim light produces a large pupil to allow as much light as possible to enter the eye. In bright light the pupil is reduced to a small size to restrict the amount of light entering.

The muscles of the **iris** can contract or relax to change the size of the pupil. The iris consists of two types of muscle – **radial** and **circular**. Radial muscles are like the spokes of a wheel moving out from the edge of the pupil through the iris, and circular muscles form rings within the iris around the pupil.

* In dim light the radial muscles contract (and the circular muscles relax) – this makes the pupil larger
* In bright light the radial muscles relax (and the circular muscles contract) – this makes the pupil smaller.

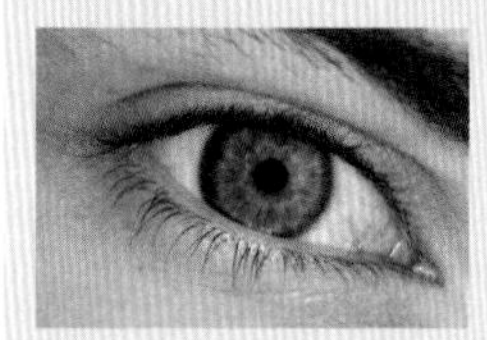
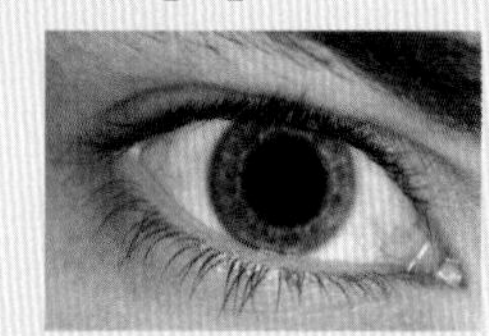

Figure 6.4 The pupil gets larger in dim light

The changing size of the pupil is an automatic reaction – we do not deliberately control it. It is called the **pupillary reflex**. (Changing the thickness of the lens is also an example of a reflex.)

▶ Voluntary and reflex actions

Most of our actions are **voluntary**. This means we deliberately choose to do them and they involve conscious thought. However, there is another type of action that does not involve conscious thought – these are **reflex actions**.

Reflex action

If you accidentally touch a very hot object you respond immediately by rapidly withdrawing your hand from the danger area. The advantage of this is that you move the hand away before it can get burned too badly. This type of action does not involve any 'thinking' time, as the time taken to consider a response would cause unnecessary damage to the body. All reflex actions have two main characteristics in common:

* they occur very rapidly
* they do not involve conscious control (thinking time).

What makes a reflex action so rapid? In a reflex pathway the total length of the nerve pathway is kept as short as it possibly can be. For example, the knee jerk reflex travels from the knee up to the base of the spinal cord and back into the leg. In addition, there are relatively few gaps between neurones (synapses), as these are the places where impulses travel relatively slowly.

Figure 6.5 shows the nerve pathway involved when a hand touches a hot object. There are three types of neurone involved in this response.

The diagram shows that both the association and motor neurones begin with the cell body (unlike the sensory neurone). The diagram also shows that only two synapses (short gaps between neurones that

slow nervous communication) are involved in this pathway. This system of structures involved in a reflex is called a **reflex arc**.

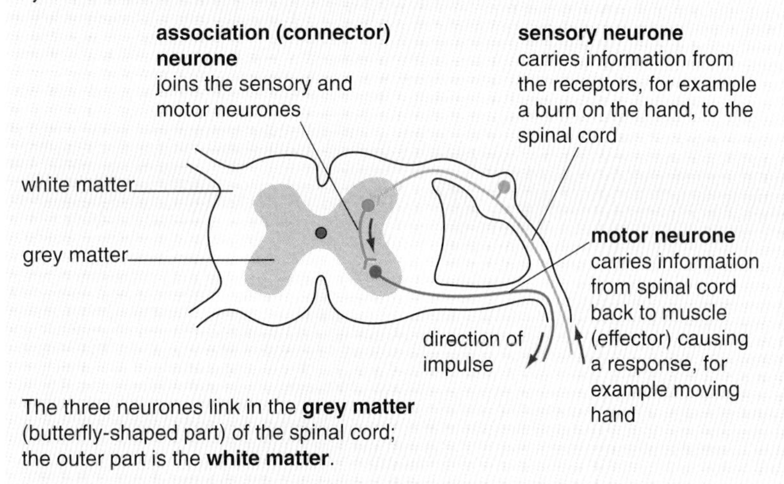

Figure 6.5 The pathway of a reflex action

Note: the pupillary reflex arc referred to earlier is different in that it remains wholly within the brain and doesn't enter the spinal cord. The reflex arc shows that neurones are specialised for their function in being very long.

Although it is not apparent in the diagram of the reflex arc they are also very specialised by having:

* branched ends that allow them to transmit impulses over a greater area or to make more connections with other neurones
* an insulating sheath that surrounds them to help speed up the transmission of an impulse.

The diagram also shows that there are small gaps called **synapses** between adjacent neurones. These act as **junctions**, with transmission across synapses caused by the diffusion of transmitter chemicals across the junction. The transmitter chemical is produced by the end of the neurone leading into the synapse and if produced in high enough concentration will trigger an impulse in the next neurone. While synapses may slow transmission they allow a greater degree of control at these junction points.

▶ Hormones

Another type of communication system is controlled by hormones. Hormones are chemicals produced by special glands that release them into the blood. Although the hormones travel all round the body in the **blood**, they only affect certain organs called **target organs**. Obviously the target organ(s) differ for each hormone and with some hormones many organs are affected.

Hormones usually act more **slowly** than the nervous system and over a longer period of time. Good examples to illustrate these points are the sex hormones, oestrogen and testosterone. The changes brought about by testosterone (males) and oestrogen (females) come about over many years.

Question

2 Draw a table to highlight the major differences between the nervous system and hormone action.

Insulin and blood glucose levels

Insulin is the hormone that prevents blood glucose (sugar) levels from becoming too high. Glucose is constantly needed by all cells for respiration and therefore must always be present at a sufficient concentration. However, if there is too much glucose in the blood this can damage the cells of the body as a result of water loss by osmosis.

Insulin is produced by special cells in the **pancreas** in response to increasing or high blood glucose levels. This usually occurs after a meal, especially if the meal is rich in carbohydrates. The insulin acts to **reduce** blood glucose levels by converting the excess glucose to **glycogen** (which is stored in the **liver**), and by increasing **respiration**.

When blood glucose levels are low, less insulin is produced. This means that the above processes (which would decrease levels even further) do not take place or take place at a slower rate.

Figure 6.6 highlights the relationship between blood glucose level and insulin.

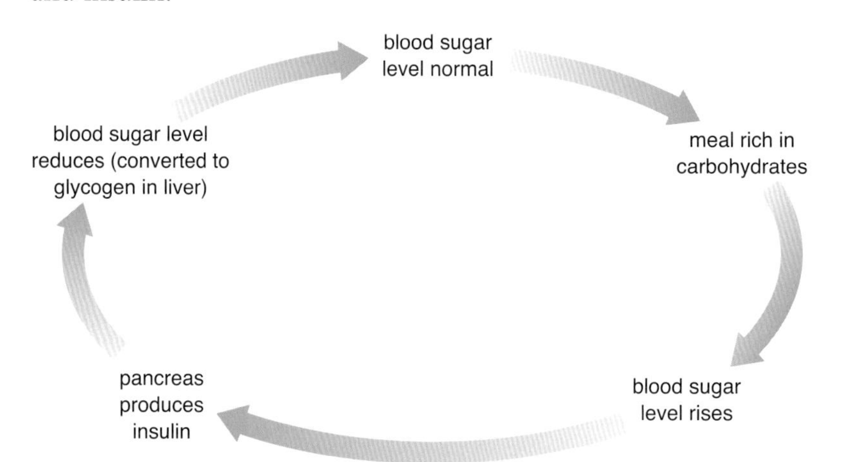

Figure 6.6 Insulin and blood sugar levels

Figure 6.7 shows how the blood glucose level typically varies after eating a meal.

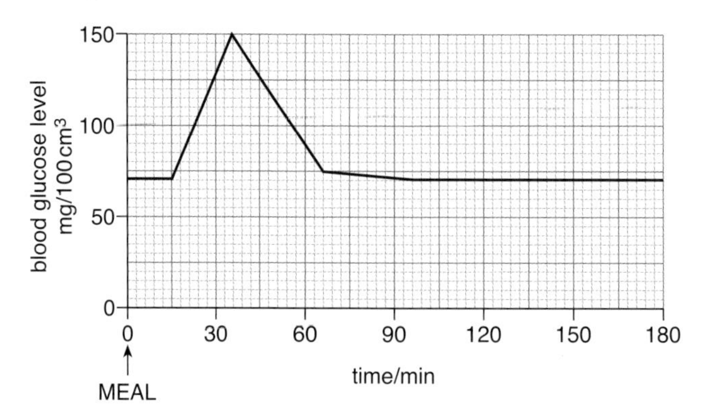

Figure 6.7 The effect of insulin on blood glucose levels

Negative feedback

Figure 6.7 shows that as the blood sugar level increases, the rate of insulin production also increases in order to reduce blood glucose levels. Similarly it is the fall in blood glucose level that slows down or stops the release of insulin. This is an example of **negative feedback**, where the level of insulin released is controlled by the level of glucose in the blood. Feedback mechanisms are an important feature of most hormones. Negative feedback is vital in the control of blood glucose because it is just as important to prevent the glucose level from falling too much as it is to prevent it from becoming too high. In general, negative feedback is a mechanism to ensure that the level of something, e.g. sugar, doesn't deviate too far from the normal value.

A second hormone is involved in controlling blood sugar level. **Glucagon** is also produced in the pancreas. It is produced when blood glucose levels are falling and it acts by reversing the changes that insulin brings about in the liver.

Question

3 Draw a flow diagram to show the process of negative feedback in blood sugar level control.

Diabetes – when blood glucose regulation fails

Diabetes is a condition where the body does not produce enough insulin to keep blood glucose levels at a normal level. Individuals that develop diabetes are unable to control their blood glucose levels without treatment and the following symptoms are often present:

* There is **glucose in the urine**. This happens because **blood glucose levels are so high** that some sugar is removed through the kidneys.
* Affected individuals are often **thirsty** and because they drink so much they need to go to the toilet a lot.
* **Lethargy** may result.

Diabetes is a fairly common (and increasing) condition in young people. It is usually treated by the injection of insulin and by a carefully controlled diet where the intake of carbohydrate is accurately monitored.

Figure 6.8 shows a young girl injecting herself with insulin. Even with the use of insulin injections and a carefully controlled diet, it is difficult for people with diabetes to control their blood sugar levels very accurately. Problems may arise if too much insulin is injected or if enough food is not eaten at regular intervals. If the blood sugar level drops too far, a hypoglycaemic attack (a hypo) may occur and unconsciousness will result. If blood glucose levels remain too high for a long period of time, serious medical complications can result.

The type of diabetes normally developed in childhood, and discussed above, is referred to as **Type 1 diabetes**. The diabetes that usually only develops in older people is **Type 2 diabetes**. It has a slightly different cause in that insulin is produced but stops working effectively. Type 2 diabetes is often associated with poor diet, obesity

Figure 6.8

Question

4 a) Suggest why people with diabetes often take extra glucose (a biscuit or a glucose drink) before they take part in vigorous exercise.

b) People with diabetes check their blood glucose levels regularly. Suggest what they could do if their blood glucose level is too high when tested.

and lack of exercise. The increasing number of people with these characteristics largely explains the increase in the number of people with diabetes. (**Note:** you do not need to know the difference between Type 1 and Type 2 diabetes.)

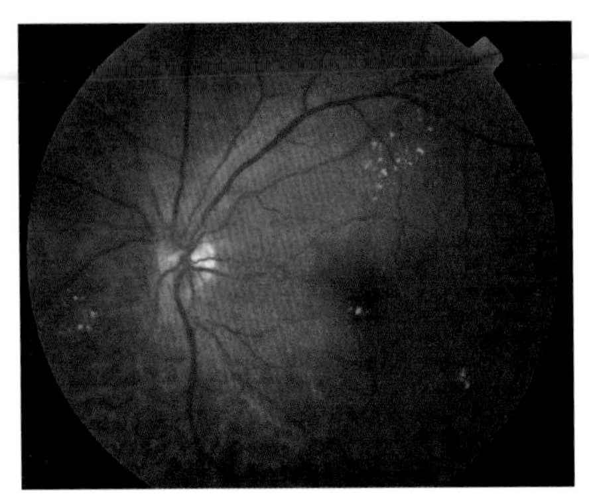

Figure 6.9 Retina of a diabetes sufferer. The areas of small yellow dots, caused by leakage from damaged blood vessels, can cause permanent loss of vision

Search ▶ diabetes

Long-term effects and future trends

People who have had diabetes for a long time (in some cases undiagnosed and unknown) and whose blood sugar level is not tightly controlled run the risk of developing serious long-term complications. These include **eye damage** or even blindness, **heart disease** and **strokes** and **kidney damage**. These complications are usually due to high blood sugar levels damaging the capillaries that supply the part of the body involved.

To find out more about diabetes visit www.diabetes.org.uk.

▶ Sensitivity in plants

Plants, like animals, respond to changes in the environment. However, they respond to fewer types of stimuli and in general the response is slower. Plants respond to the environmental stimuli that have the greatest effect on their growth. Roots grow towards water when a moisture gradient exists and shoots tend to grow away from the effects of gravity, i.e. they grow upwards. Reasons for these responses are fairly obvious as they ensure that the plants react in such a way that they receive the best conditions for growth.

The response of a plant shoot to light is called **phototropism** and this response has been investigated in detail to establish how it occurs.

▶ Phototropism – responding to light

Most of you will have observed that plants grow in the direction of a light source. Plants left on a window-sill or against the wall of a house usually do not grow straight up, but bend towards the light source. You can probably also guess that this response ensures that the plant stem and leaves receive more light than they otherwise would do if there was no such response. This means that *more* photosynthesis takes place and there will be *more* growth.

Although it is easy to observe the effect of phototropism, what causes it to occur? Figure 6.10 shows the growth of young seedlings in unilateral light (light coming from one side or source only) and highlights some of the features of phototropism. Can you use the diagram to identify what part of the plant perceives (is sensitive to) the light source?

Figure 6.10 shows that it is the shoot tip that is sensitive to light, as when it is covered the phototropic response does not occur. Plant stems produce a hormone called **auxin** in the tip. When a stem is illuminated from one side this hormone tends to accumulate more on the non-illuminated side.

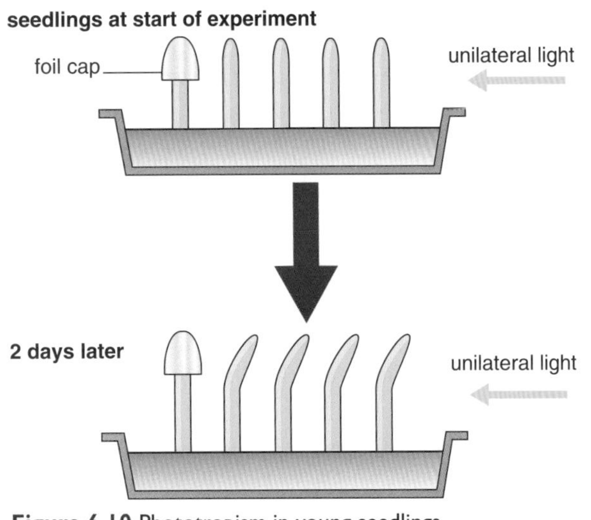

Figure 6.10 Phototropism in young seedlings

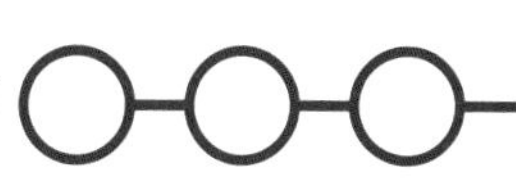

Question

5 Using your knowledge of phototropism, explain the following:
   a) How is it possible to keep plants in unilateral light, for example a plant growing on a window-sill, growing straight?
   b) Why do plants that have had their tips removed often stop growing?

As the effect of the hormone is increased growth this leads to the non-illuminated side growing more rapidly than the side that is receiving most light. The differential growth that occurs when one side of the stem grows more than the other side leads to the stem bending in the direction of the light.

The auxin is produced at the tip of the shoot and accumulates on the shaded side as it diffuses down the shoot. The bending is caused by the auxin promoting greater cell elongation in regions where it is in high concentration.

▶ The commercial use of plant hormones in weed control, flowering and fruit formation

Auxin is only one of a range of plant hormones that have been identified. Many plant hormones are used by horticulturalists and gardeners for controlling weeds and in increasing flower and fruit production.

Hormones as weedkillers

It is often important to eliminate weeds as they can affect crop production by using the resources that the crops need. This competition for water, minerals, light and space between weeds and crops can severely limit the growth of crops.

A range of hormones can be used in weedkillers, but most contain synthetic (artificially produced) auxin that is similar in structure and effect to the naturally occurring auxins in plants. These weedkillers are particularly effective as they are selective. They are effective against broadleaved plants such as daisies and plantain, but do not have an effect on narrow leaved plants such as grasses and cereals.

They work by causing the weeds to grow in such a rapid and uncontrolled way that they disintegrate.

Question

6 Why are gardeners advised not to spray weedkiller if heavy rain is expected?

Figure 6.11 A dock plant before (left) and after (right) treatment with a weedkiller

The stimulation of flowering and fruit formation

Hormones can be sprayed onto commercially grown flowers, crop plants and fruit trees to ensure that flower production and/or fruit production is maximised.

Examples of this in action can be seen in fruit growing regions such as County Armagh. Normally when flowers are pollinated and fertilised these processes set in train a series of hormonal changes that leads naturally to fruit production. However, poor weather that damages flowers and decreases the number of insects available for pollination often leads to low pollination rates. If nature is left to run its course the fruit production would also be low. The use of hormones acts as a substitute for the pollination and fertilisation processes and ensures that the fruit production is not limited by poor pollination rates. The use of artificial hormones can also help make all the fruit develop and ripen at the same time. This can lead to greater efficiency at the time of harvesting.

Figure 6.12 These grapes are ready to eat. If they are seedless, artificially added hormones will have been used in their development

There are many examples where hormones are used to produce a fruit that is more attractive to the consumer. Seedless grapes are becoming increasingly popular. Naturally an ovary in a vine will only develop into a grape if its ovules are fertilised to become seeds. The use of synthetic hormones can deliberately bypass and prevent the pollination and fertilisation process, ensuring that the grapes develop without seeds.

Hormones in rooting powder and tissue culture

Cuttings of geraniums and other plants can be taken to grow new plants. The cutting is taken and planted in compost and eventually roots will develop and a new plant forms. The process can be accelerated by dipping the cut stems in rooting powder

containing hormones that stimulate root growth.

The process of tissue culture involves taking very small sections of plant tissue (callus) and placing them on agar or another nutrient-containing medium. Hormones are added to encourage the small section to differentiate and develop into new plants.

Much of the knowledge gained about the nervous system and hormones shows some general principles about scientific research:

* it is often developed in stages using different lines of evidence
* it involves collaboration among different scientists
* it is validated by peer review – the process where other scientists of equivalent status review and comment on new research.

▶ Exam questions

1 The diagram shows a section through an eye.

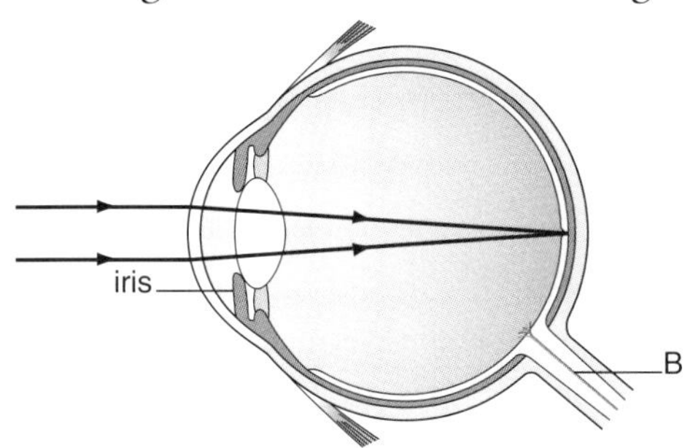

a) What is the function of the iris? *(1 mark)*

b) Name part B. *(1 mark)*

c) Use the diagram and your knowledge to explain how light rays that enter the eye are focused onto the retina. *(1 mark)*

2 The table shows reaction times in response to the stimulus of a light being switched on.

Age/years	Reaction time/milliseconds	
	Males	Females
20	240	320
30	220	260
40	260	340
50	270	360
60	380	440

Based on data in the above table, the percentage increase in reaction times between ages 20 and 60 was calculated.

Percentage increase in reaction times between ages 20 and 60	
Males	Females
58%	38%

a) Use the above data to compare the reaction times of males and females as they age. *(2 marks)*

The diagram shows a reflex arc.

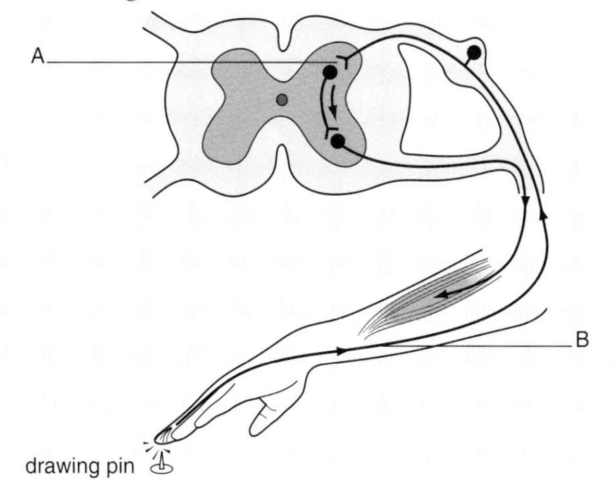

b) i) Name part B. *(1 mark)*

ii) Describe how B is adapted to its function. *(2 marks)*

c) A is a synapse. How does the nerve impulse pass across this synapse? *(2 marks)*

d) Explain why the response to the drawing pin through the reflex arc would be faster than that to a light being switched on. *(2 marks)*

3 The graph shows the changes in the blood glucose level of two boys, Patrick and Glen. Both have fasted (had no food intake) for 12 hours before taking an energy drink at time 0.

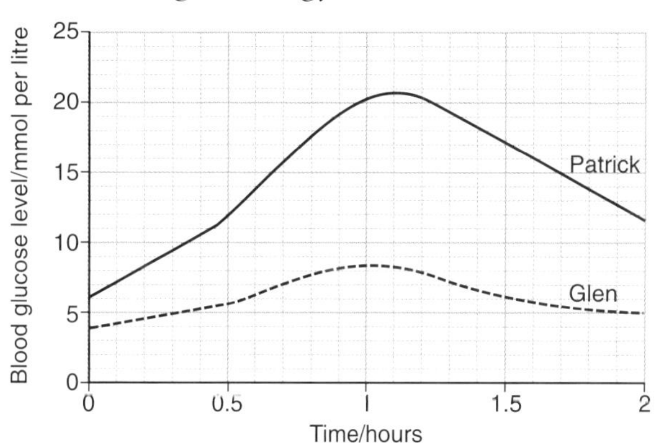

a) Compare the changes that take place in Patrick's blood glucose levels with those of Glen. Use information in the graph to support your answer.

In this question you will be assessed on your written communication skills including the use of specialist science terms. *(6 marks)*

b) Explain the changes taking place in Glen's blood glucose levels. *(2 marks)*

c) Blood glucose levels above 10 mmol per litre result in glucose appearing in the urine, a symptom of diabetes.

i) Use evidence from the graph to identify which of the boys has diabetes. *(1 mark)*

ii) State another symptom of diabetes that this boy might have. *(1 mark)*

4 The diagram shows part of the mechanism that controls blood glucose concentration.

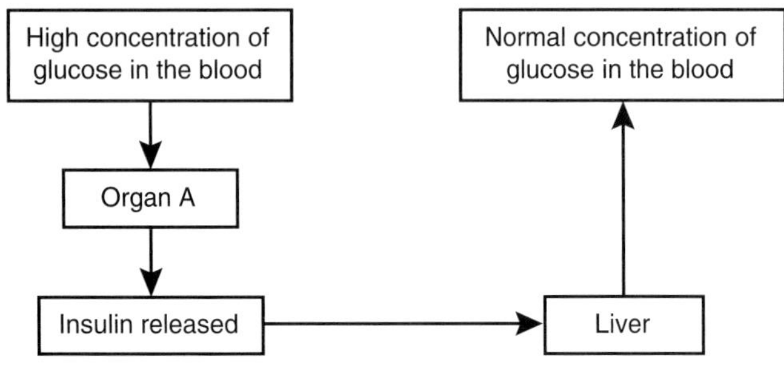

a) Name organ A. *(1 mark)*

b) Explain why the blood glucose concentration becomes high after eating a meal. *(1 mark)*

c) Describe how insulin reaches the liver. *(1 mark)*

d) Explain how insulin causes the liver to reduce the blood glucose concentration. *(3 marks)*

e) Use the information in the diagram to help explain how the control of blood glucose concentration involves a feedback mechanism. *(3 marks)*

f) Some people are unable to control their blood glucose concentration. Name this condition. *(1 mark)*

5 An experiment was carried out to investigate the effect of light shining from one side on a plant shoot.

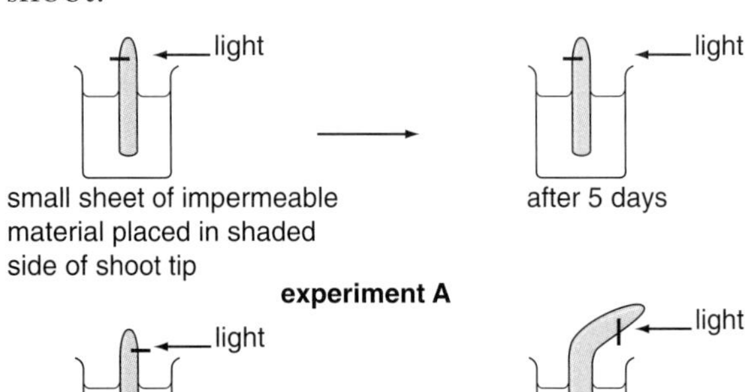

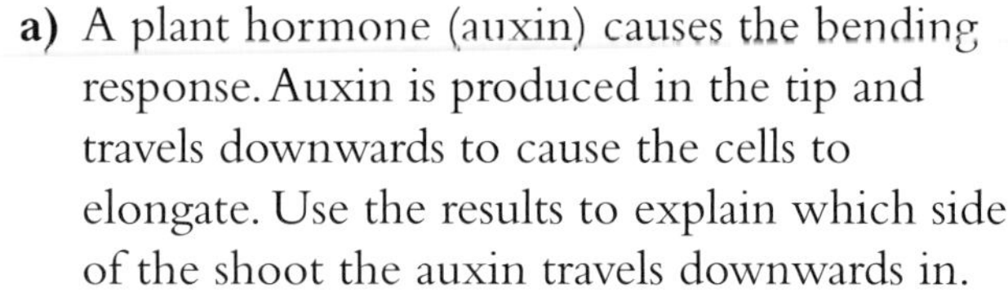

a) A plant hormone (auxin) causes the bending response. Auxin is produced in the tip and travels downwards to cause the cells to elongate. Use the results to explain which side of the shoot the auxin travels downwards in. *(2 marks)*

b) i) Name the response shown by the shoot in experiment B. *(1 mark)*

ii) Explain how this response will benefit the plant. *(2 marks)*

6 The graph shows the relationship between plant hormone levels and the percentage of apples to fall off the trees before harvesting.

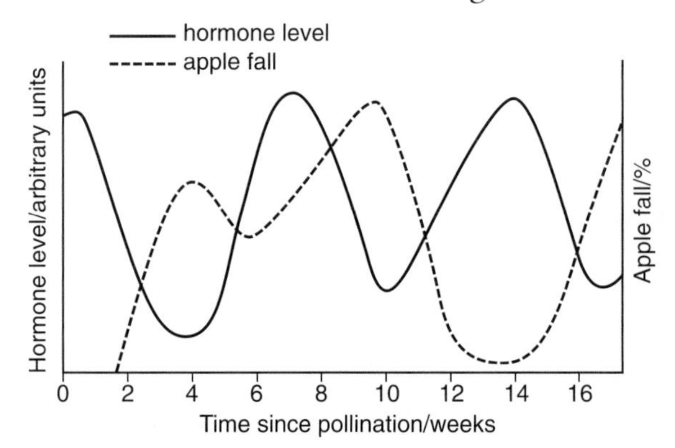

a) Describe the relationship between hormone level and the percentage of apples that fall. *(1 mark)*

b) Suggest **one** reason why farmers spray apple trees with artificial plant hormone. *(1 mark)*

c) Give **two** other commercial applications of plant hormones. *(2 marks)*

7 Ecological Relationships and Energy Flow

Ecology is the study of habitats and their environments. Most GCSE students will have an opportunity to complete at least one ecological study where they will sample the numbers of different types of organisms in an area and investigate some of the environmental factors that explain their distribution.

Realistically, when investigating the number of plants and animals in a habitat such as a woodland, shoreline or sand dune system, it will not be practical to count all the living organisms and investigate every square metre. Instead the area is **sampled** – a number of sub-sections within the habitat are sampled to give an overall picture.

▶ Sampling plant distribution

The most common way of sampling plant distribution is to use a **quadrat**. A quadrat is a square frame. Some quadrats have 1 m sides giving a quadrat area of 1 m^2; others have 50 cm sides giving an area of 0.25 m^2.

When investigating plant distribution, **percentage cover** is often measured. This is the percentage of the quadrat covered by a particular type of plant. It is difficult to estimate exactly what percentage of a quadrat is covered by a particular type of plant, so it is normal to round up to the nearest 10%. (An exception is if there are any plants with a percentage cover between 1 and 5 – this is recorded as 1 and not as 0.) Therefore, when sampling for percentage cover using a quadrat, the possible values obtained are 0, 1, 10, 20 and so on up to 100%. Figure 7.1 shows the estimated percentage cover of three plants in a quadrat.

Figure 7.1 Using a quadrat to measure percentage plant cover

Obviously, in the area being sampled one quadrat would not give very **reliable** results. To gain reliable results it is important to take values from a number of quadrats (the number selected will depend on the size of area to be sampled but for a reasonably large area it is usually necessary to

Questions

1. Suggest why the quadrats must be placed randomly in the habitat.
2. Suggest why it is better to use percentage cover, rather than number, when sampling many types of plants.

Figure 7.2 Zonation on a rocky shore

use 20 or more quadrats). It is also very important that the quadrats are **randomly** placed over the habitat. This can be done by sub-dividing the habitat into a grid and using random numbers to generate co-ordinates.

When the data for all the quadrats is collected, the values for each type of plant can be averaged to give an overall estimate of percentage cover for the habitat as a whole.

For some plants it is not necessary to use percentage cover – counting the number of plants present is suitable.

In a habitat that shows clear gradation over an area, for example moving from low tide line to high tide line on a rocky shore, randomly placed quadrats will not identify the gradation with distance up the shore.

Figure 7.2 shows a rocky shore. It is possible to see from the photograph how the types of plants change from the bottom of the shore to the top.

To investigate how the shore changes a **belt transect** could be used. In a belt transect, quadrats are placed along a line from the bottom of the shore to the top. Quadrats can be used continuously or at intervals along the line depending on the distance involved.

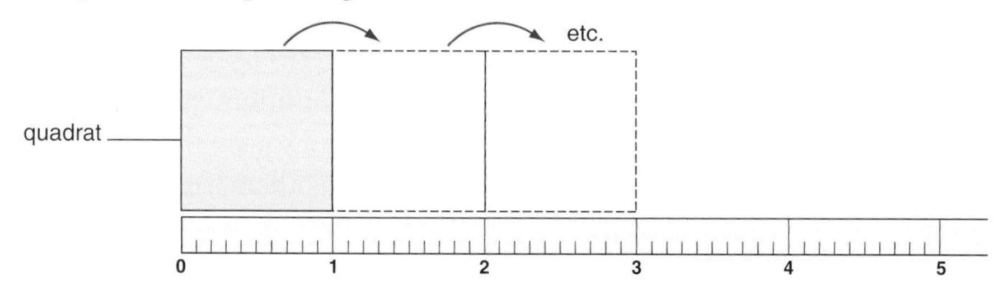

Figure 7.3 A belt transect

▶ Sampling animal distribution

Some animals can be sampled using quadrats. However, this is only useful for animals that don't move or only move very slowly, such as limpets on a rocky shore.

Most animals do move so they need to be trapped when sampling. When sampling animals, percentage cover is not normally used and the number of animals is recorded instead. There are a number of simple methods that can be used for sampling animals in a habitat.

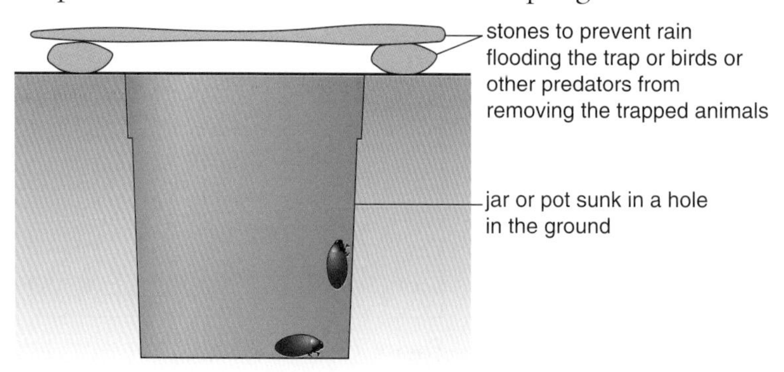

Figure 7.4 The pitfall trap

A **pitfall trap** is used to trap small ground-dwelling animals such as beetles that fall into the trap as they move along the ground.

Question

3 Suggest why a pitfall trap is not suitable for sampling butterflies.

Nets are very useful for trapping insects and other animals that live in long grass in a meadow or in a river or pond. They are often particularly effective when moved in large sweeps through the grass or water; this is why they are sometimes referred to as sweep nets. **Pooters** can be used for sucking up and collecting small insects.

Figure 7.5 A net

▶ Investigating abiotic factors

Most ecological investigations involve the analysis of some of the **abiotic** (non-living) factors that could affect the distribution of plants and animals. Examples of the abiotic factors that can be investigated include:

* **Wind** – wind speed can be analysed using anemometers. Wind speed can be very important, affecting the numbers and distribution of plants and animals in exposed habitats such as sand dune systems.
* **Water** – soil moisture levels can be calculated by taking soil samples and weighing them to find their mass. The soil samples are then dried in an oven until completely dry and reweighed. The difference in mass as a percentage of the original mass gives a value for percentage soil moisture. Soil moisture levels can be an important factor in the distribution of many plants and animals.
* **pH** – pH can be measured using soil testing kits or probes. The pH of a soil is very important in the distribution of many plants. Some plants will only grow in relatively acidic soils, e.g. heathers, and some will only grow in relatively alkaline soils, e.g. some orchids, but most prefer soil pH to be around neutral.
* **Light** – light can be measured using light meters. Light is particularly important in the distribution of plants. While all plants need light to photosynthesise, some need high light levels to thrive whereas others can survive in very low light levels.
* **Temperature** – temperature can be measured using a thermometer.

Biotic factors are living features of the environment. **Biodiversity** is a measure of the number of different types of plant and animal **species** in an area.

The following case study gives an example of abiotic and biotic data in a habitat.

▶ Case study – plant distribution in a sand dune system

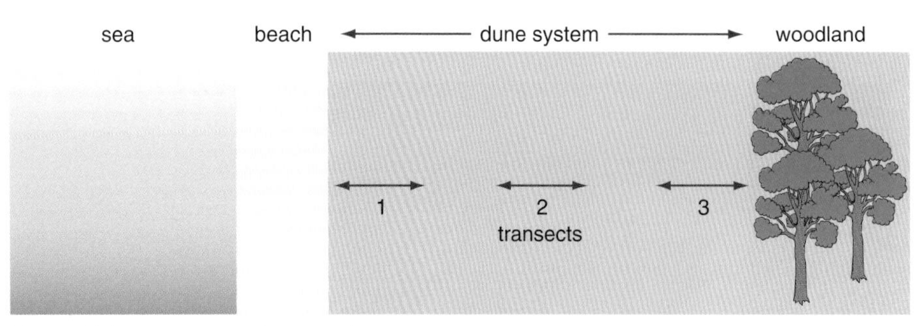

Figure 7.6 Investigating plant distribution in a sand dune system

The following data was collected from a survey in a sand dune system. As the sand dune system extended for over 1 kilometre the data could not be collected from a continuous belt transect but was collected from three 'interrupted' belt transects. The first belt transect was from the start of the first dune nearest the shore and extended inland. The third belt transect was at the very end of the dune system, just before the typical dune vegetation was replaced by woodland. The second transect was halfway between the other two. Data was collected from 20 quadrats in each belt transect.

Figure 7.7 shows some of the abiotic information gathered and the distribution of three plants typical of sand dunes. The appropriate abiotic data was collected for each quadrat sampled in each transect.

Marram grass is the grass typical of sand dunes and has an important role in binding the sand within the dunes together and enabling the dunes to become stabilised. Unlike most other plants it can grow in very unstable conditions such as those found near the shore. Heather, a small shrub, grows where the soil becomes more stable and moist. The larger woody shrub gorse will thrive when the soil becomes even more stable and contains more nutrients and moisture.

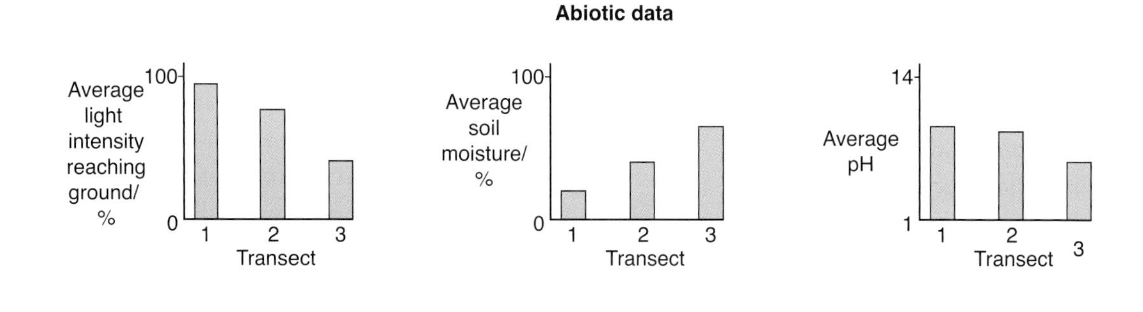

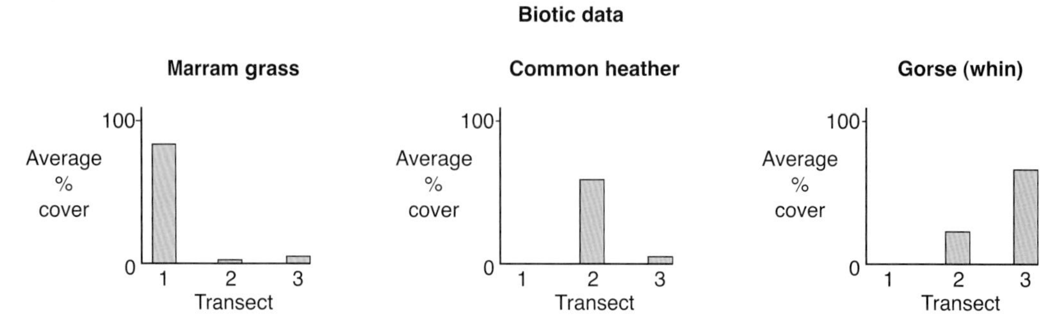

Figure 7.7 Abiotic data and plant distribution

The graphs show that marram is very common in Transect 1 only. Heather is not found in Transect 1 but is the most common plant found in Transect 2. Gorse is the most common species found in Transect 3 but is very uncommon or not found at all in Transects 1 and 2.

You should be able to interpret this data and give possible explanations for the plant distributions based on the information given.

Possible explanation

The marram grass is able to grow near the shore and is important in the formation and stabilisation of dunes (you have been given this information). It is also logical to conclude that it requires high light levels to grow and possibly grows best at a slightly alkaline pH (from the abiotic data). However, it appears to be much less successful further inland where the soil becomes more stable, more moist and more acidic. It appears that the changing conditions further inland favour other plants such as heather and gorse and that these out-compete the marram grass. The low light level in Transect 3 (created by shade from the gorse shrubs) probably prevents the marram grass from growing.

The small shrub heather gains a foothold in the more stable Transect 2 but cannot compete against the larger woody shrubs in Transect 3. Like the marram grass, the heather is probably unable to survive in the shade of the larger gorse.

It is important to note that plants and animals can influence the abiotic data – it is not only the other way around. In this example, the ground light levels in each transect are entirely controlled by the shade produced by the plants growing there.

Question

4 Based on the information provided, give one reason why the data for this investigation could be considered reliable. What could be done to further increase the reliability?

▶ Validity and reliability

During this course you have come across the term **reliability** on a number of occasions. If data are reliable, then if someone else was to repeat the investigation they would get similar results. Reliability can be increased in experiments by doing repeats or taking many measurements.

The validity of information provided is an indication of whether you are actually able to draw conclusions from the information. The concept of a 'fair test' applies as much to ecological experiments as it does to laboratory-based ones. For example, you will only be able to state categorically that a particular abiotic factor is responsible for a change in vegetation if other abiotic factors are controlled or similar.

The **distribution of animals** is often affected by other factors such as the availability of **food**, **mates** and the establishment of **territories**.

Question

5 a) Using the information opposite suggest an explanation for the increase in the world human population.

b) Explain how the increasing human population size harms the environment:

i) locally ii) globally.

▶ Changes in population

Population numbers change over time. There are many factors that can contribute to population change but they can be summarised by the following equation:

$$\text{population growth} = (\text{birth rate} + \text{immigration}) - (\text{death rate} + \text{emigration})$$

There are some other key terms you need to know. These include:

* **Habitat** – the place where an organism lives.
* **Population** – the number of one type of organism (species) in an area.
* **Community** – the total number of organisms from all the populations in an area.
* **Environment** – the factors, both physical (abiotic) and living (biotic), that affect organisms in a habitat.
* **Ecosystem** – the community of organisms that are interdependent on each other and the environment in which they live.

▶ Classification

You should be able to use keys to identify organisms and classify them into major groups. This is particularly important during fieldwork as you may come across some organisms you do not recognise.

In addition to identification, classification is also important for:

* studying how organisms have changed over time
* studying biodiversity
* the conservation of species.

This includes identifying the main groups as listed in the following table.

Group	Nutrition	Cell wall	Cellular organisation
Protoctista	Saprophytic or photosynthetic	Cellulose cell wall or none	Single celled with nucleus or algae that are not truly multicellular
Bacteria	Saprophytic	Non-cellulose	Single celled with no nucleus
Fungi	Saprophytic or parasitic	Non-cellulose	Single or multicellular – can be 'acellular' with it being difficult to distinguish individual cells and nuclei scattered throughout the organism
Plants	Photosynthesis	Cellulose	Single or multicellular – 'typical' cell arrangement with a nucleus
Animals	Eating organic food	None	Single or multicellular – 'typical' cell arrangement with a nucleus

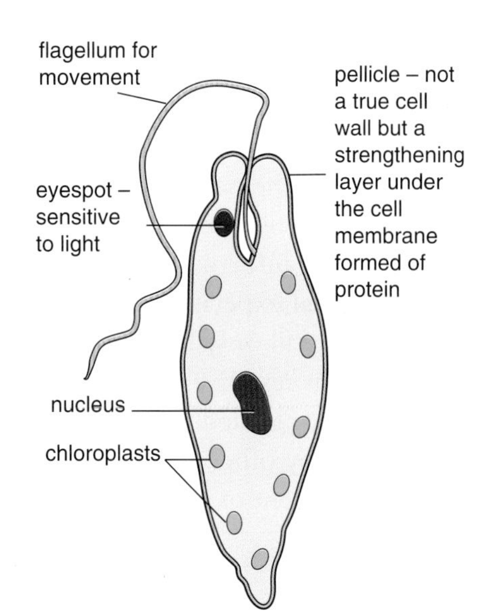

Figure 7.8 Euglena

Note 1: Saprophytic nutrition usually involves releasing enzymes into the substrate (surrounding medium, e.g. soil) and absorbing the digested products of the enzyme action.

Note 2: Some organisms are very difficult to classify. These include e.g. *Euglena*, which has both animal and plant characteristics – do we include it as a plant or an animal? This is why the above table classifies single-celled plants and animals in a separate group called the Protoctista.

Note 3: It is also sometimes difficult to identify which species an organism belongs to or where one species merges into another. A species is a group of organisms, with shared features, which can breed together to produce fertile offspring.

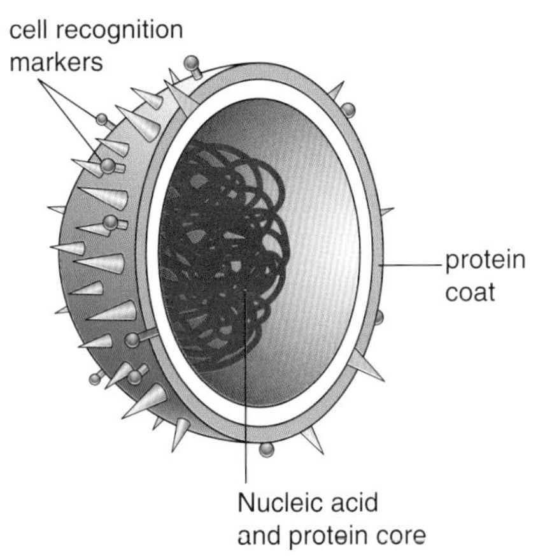

Figure 7.9 A virus

Note 4: Viruses are a very complex group. Viruses, e.g. the HIV virus that can cause AIDS, lack proper cellular organisation. They have a DNA/RNA (nucleic acids – the building blocks of chromosomes) core and an outer protein coat without the typical cytoplasm of other cells. They can only live when they gain access to other cells and many biologists therefore regard them as non-living. If many regard them as non-living where do they fit into a classification system of living organisms? It is important to note that classification systems change over time.

Group	Features
Flowering plants	True roots, stems and leaves with specialised vascular tissue; flower and seed production with a fruit, e.g. bluebell, horse chestnut
Annelids	Segmented worms with chaetae for grip; body temperature not constant, e.g. earthworm
Insects	Have an exoskeleton and three parts to the body (head, thorax and abdomen); three pairs of jointed legs, two pairs of wings; body temperature not constant, e.g. dragonfly, mosquito
Chordates	Animals with backbones, e.g. fish, reptiles and mammals

▶ The transfer of energy and nutrients

Examples of ecosystems include grasslands, woodlands and lakes. If ecosystems are able to remain stable for long periods of time there must be some way in which energy continually enters the system to replace the energy that is lost through respiration and the many energy-requiring activities that occur. Where does this energy come from?

The energy comes from the Sun and is trapped by green plants in the process of photosynthesis. Plants that can photosynthesise are known as **producers** as they produce their own food and they in turn provide food and energy for other organisms. The herbivores (plant-eating animals) that feed on plants are known as **primary consumers** and the carnivores (animals that eat other animals) that feed on primary consumers are known as **secondary consumers**. Animals that feed on secondary consumers are **tertiary consumers** and so on.

This sequence of producers trapping the Sun's energy and this energy then passing into other organisms as they feed, is known as **energy flow**.

The different stages in the feeding sequence can also be referred to as **trophic levels**. Producers occur in trophic level 1 and primary consumers are trophic level 2, etc.

▶ Food chains and food webs

Figure 7.10 shows a sequence or chain of living organisms through which energy passes. It is an example of a **food chain**. Food chains show the feeding relationships and energy transfer between a number of organisms. Examples of some other food chains are shown in Figure 7.11.

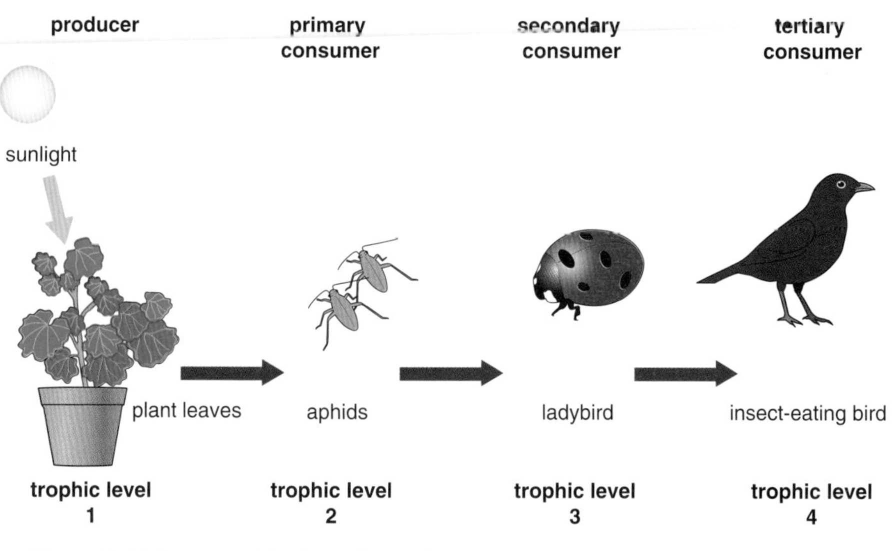

Figure 7.10 Energy and feeding relationships

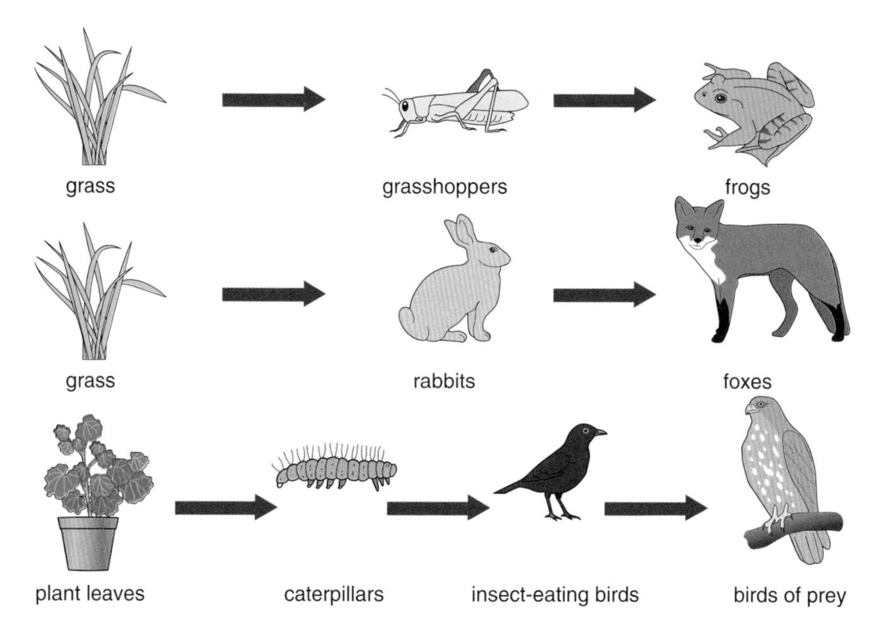

Figure 7.11 Some common food chains

Question

6 Why is it risky for animals to have only one food source?

These examples show that in all food chains the first organism is the producer and it provides food and energy for primary consumers and so on. Of course, food chains are very simplistic in that they do not show the complex interactions that usually exist. In reality very few animals have only one food source.

Food webs show how a number of food chains are interlinked and they give a much more realistic picture. Figure 7.12 shows how the food chains above are built up into a food web. The food web shown is only part of the story as there will be many more links and organisms involved than those listed.

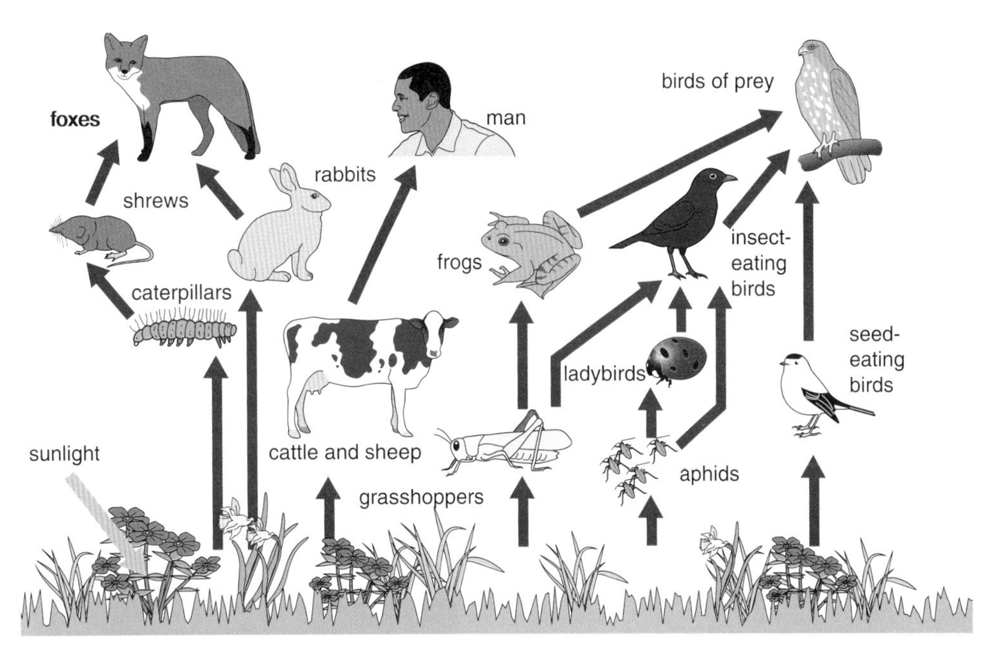

Figure 7.12 A grassland food web

▶ Pyramids of numbers and biomass

For a food web to be sustainable there must be enough food for all the organisms involved. There will usually be more producers than there are primary consumers and more primary consumers than there are secondary consumers. The number of organisms at each stage of a food chain can be represented in a **pyramid of numbers**. Figure 7.13 shows a typical pyramid of numbers. The term 'pyramid' is used as the shape will usually resemble a pyramid.

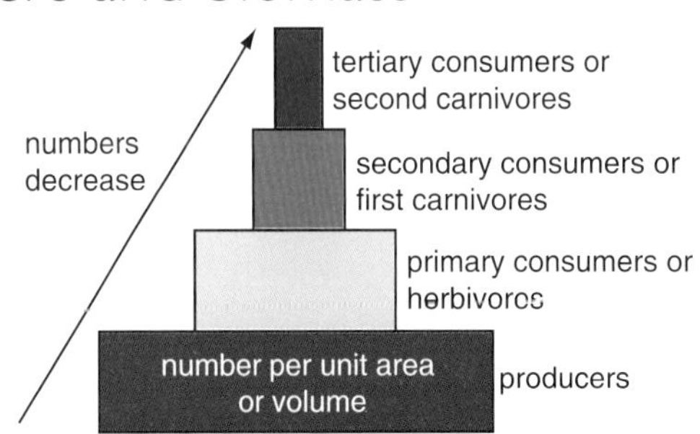

Figure 7.13 Pyramid of numbers

Pyramids of numbers can sometimes be misleading as they do not take into account the size of the organisms involved. The pyramid of numbers in Figure 7.14 highlights this problem. In the second example, each tree (the producer), may sustain many primary consumers therefore giving an atypical shape or inverted appearance.

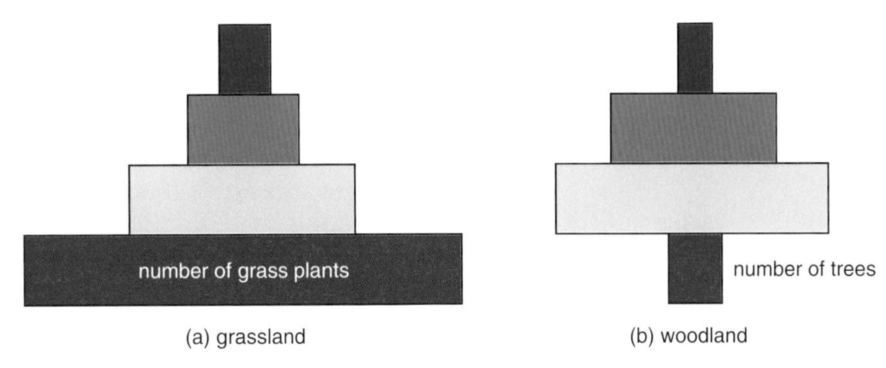

Figure 7.14 Pyramid of numbers for (a) a grassland community and (b) a woodland community

You can see this for yourself if you examine the leaves of a tree carefully. You will find many small insects feeding on, and occasionally inside, the leaves of one tree.

When looking at energy flow through a food chain it is sometimes more accurate to use a **pyramid of biomass**. These diagrams represent the mass of living tissue in the organisms concerned. Figure 7.15 shows that if we use a pyramid of biomass for the woodland pyramid in Figure 7.14 it is no longer inverted and it now has the typical pyramid shape.

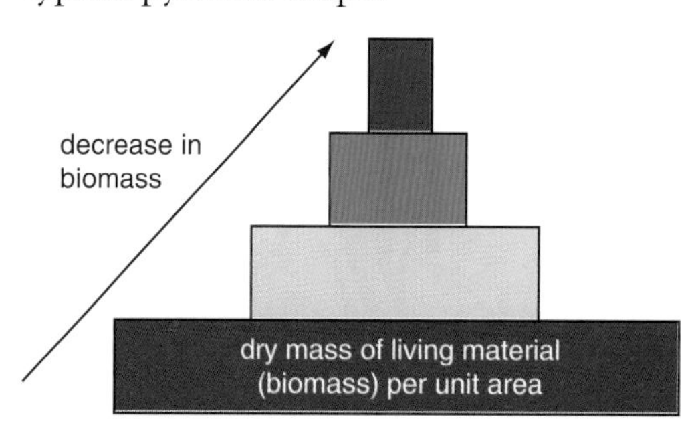

Figure 7.15 Pyramid of biomass for a woodland community

You should be able to explain the advantages and disadvantages of using each type of pyramid and also recognise the difficulties in producing pyramids if organisms feed at two different trophic levels.

Energy loss and trophic levels

All the food chains in Figure 7.11 are relatively short, containing no more than four organisms. This is because energy is lost at each stage of energy transfer. Even the absorption of light energy by plants is not particularly efficient – energy is lost as light is reflected or passes through leaves and misses the chloroplasts, or for many other reasons. However, this loss of energy is not significant as there is no shortage of light energy coming from the Sun.

The transfer of energy between plants and animals and between animals of different trophic levels is usually 10–20%. This means that for every 100 g of plant material available, only between 10 and 20 g is built up as tissue in the herbivore's body. The same applies when carnivores eat herbivores. This loss of energy is due to three main reasons:

* Not all the available food is eaten. Most carnivores do not eat the skeleton or fur of their prey, for example.
* Not all the food is digested; some is lost as faeces in egestion.
* A lot of energy is lost as heat in respiration. Respiration provides the energy for movement, growth, reproduction, etc. Heat is produced as a by-product of respiration. The heat is lost and cannot be passed on to the next trophic level.

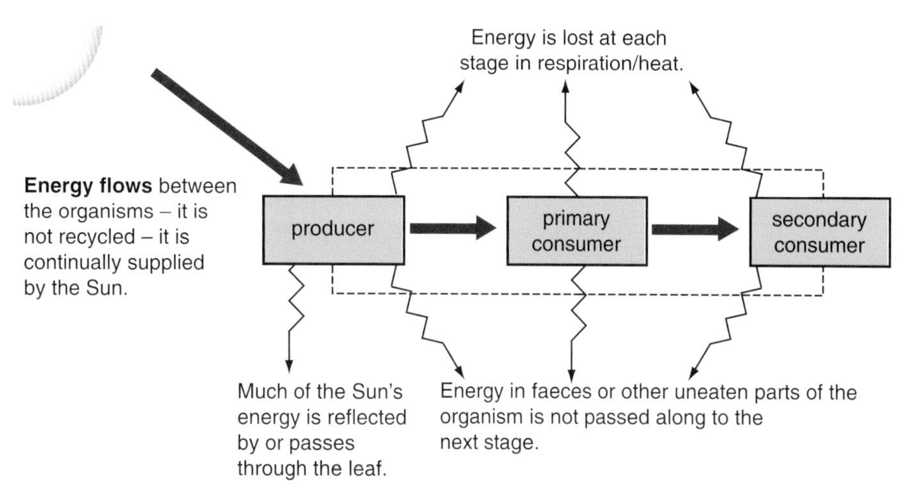

Figure 7.16 How energy is lost in a food chain

Question

7 Explain why eating rice as a staple diet is more efficient than eating meat.

Figure 7.16 shows why food chains are relatively short. So much energy is lost at each trophic level that a long food chain would not be sustainable. In effect, short food chains are much more efficient as there are fewer 'steps' and therefore less stages at which energy can be lost.

▶ Nutrient cycles

We have already noted that energy flows through food webs and that this usually happens as part of the feeding process. Figure 7.16 shows that energy must continually enter the system from sunlight as it is lost from all living organisms during the process of respiration. This is why we use the term **energy flow**.

If we look at the flow of **nutrients** in more detail we see that it differs from the flow of energy in important ways. In a stable ecosystem the overall gain or loss of nutrients from the system will be small and, unlike energy, the nutrients can be recycled as part of a **nutrient cycle**.

Nutrient cycling involves the processes of decay and decomposition. For recycling to take place, dead organisms must first be broken down during the decay process. Many organisms are involved in this process including earthworms, woodlice and various types of insect. **Fungi** and **bacteria** are the **decomposers** that break down the organic compounds into their simplest components.

Questions

8 Decomposers carry out extracellular digestion. What is meant by extracellular digestion?

9 Give two reasons why large, flat tropical leaves will decompose much more quickly than Norwegian pine needles.

The saprophytic bacteria and fungi **secrete enzymes** into the soil or dead organism. The enzymes break down the organic material and it is then absorbed by the bacteria or fungi. **Humus** is the organic content of the soil formed from decomposing plant and animal material. Decomposition takes place more quickly when conditions are optimum. These include:

* a warm temperature
* adequate moisture
* a large surface area in the decomposing organism.

The carbon cycle

The **carbon cycle** is an example of a very important nutrient cycle. Carbon is an essential element in every living organism. For example, proteins, carbohydrates and fats all contain carbon. The carbon cycle involves the exchange of carbon between living organisms, but also includes transfer between these organisms and carbon dioxide in the atmosphere. The main processes in the carbon cycle are:

* **Photosynthesis** – carbon dioxide is taken in by plants and built up into sugar and starch and other organic compounds.
* **Feeding** – animals eat the plants (or other animals) and the carbon is built up into other organic compounds that can be transferred further along the food chain.
* **Respiration** – when plants, animals and decomposers respire they return carbon compounds to the atmosphere as carbon dioxide (a form of **excretion**).
* **Decomposition** – carbon compounds in dead organisms and from **egestion** (e.g. faeces) are broken down into simpler products. As the decomposers break them down they respire and release carbon dioxide into the atmosphere.
* **Combustion** – when carbon-rich reserves of coal, oil and gas are burned the carbon is returned to the atmosphere as carbon dioxide. These fossil fuels were formed millions of years ago when plants (and animals) did not decay but were preserved, in a process called **fossilisation**, due to the particular conditions at the time.

Question

10 Use the information opposite to draw a carbon cycle.

Revisiting the carbon cycle – global warming

There is increasing evidence that the level of **carbon dioxide** in the atmosphere is rising. There is also evidence that humans are responsible. Figure 7.17 shows the **carbon cycle**.

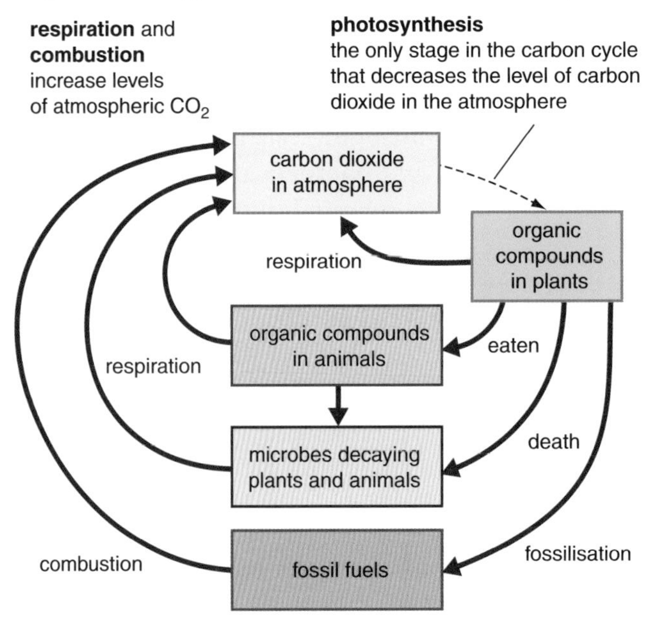

Figure 7.17 The carbon cycle

Over the last 150 years or so there have been two major changes to the way this cycle works.

* Increased **combustion of fossil fuels** has added more carbon dioxide to the atmosphere.
* Increased **deforestation** has removed many forests, meaning that less carbon dioxide can be taken out of the atmosphere by photosynthesis.

These changes mean that the carbon cycle has become **unbalanced**, leading to an *increase* of carbon in the atmosphere.

The link between increased carbon dioxide levels and global warming

It is known that carbon dioxide and some other gases in the atmosphere form a 'greenhouse blanket', trapping the heat from the sun's rays within the atmosphere. This is explained in more detail in Figure 7.18. It is thought that the increase in carbon dioxide is increasing the greenhouse effect and this is leading to global warming.

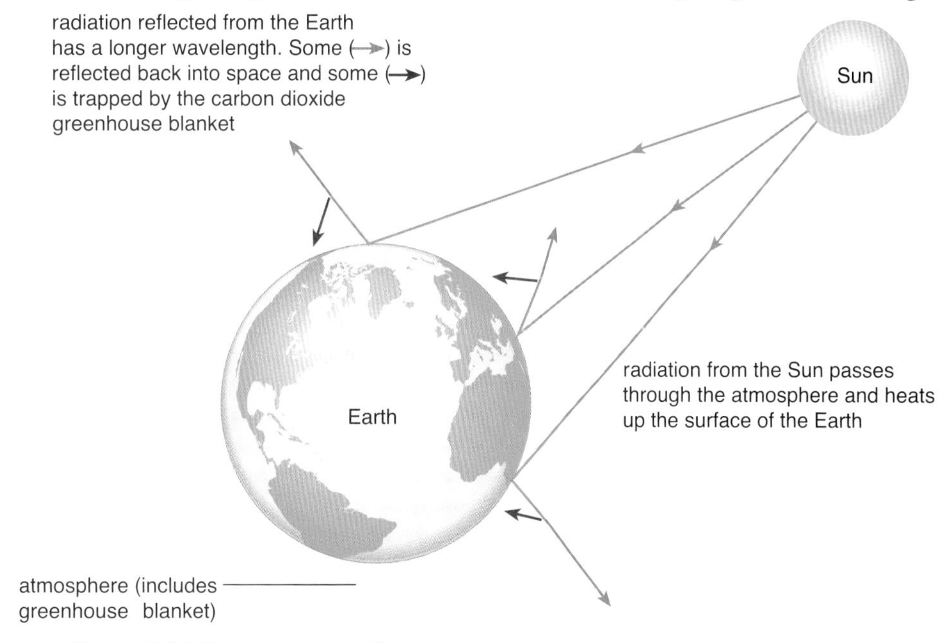

Figure 7.18 The greenhouse effect

The evidence for global warming

It is only recently that many politicians and people have accepted that it is the increase in carbon dioxide levels that causes global warming. This may be because accepting this link also means accepting that humans are responsible and therefore it is up to us to do something about it!

The effects of global warming

The warming of the atmosphere causes:

* climate change – more weather extremes such as droughts and severe storms
* polar ice-caps to melt
* increased flooding
* more land to become desert.

What can be done to reduce global warming?

* Plant more trees.
* Reduce deforestation.
* Burn less fossil fuels by using alternative fuels and/or becoming more energy efficient.

Many people believe that it is very important to act now before it is too late. We may not be able to stop global warming but we can slow it down. We can also manage it better, for example, it would be better not to build houses on areas that are likely to flood, unless we make sure there are better flood defences.

You should be aware that there is still considerable controversy surrounding global warming – the evidence, causes and possible solutions.

The nitrogen cycle

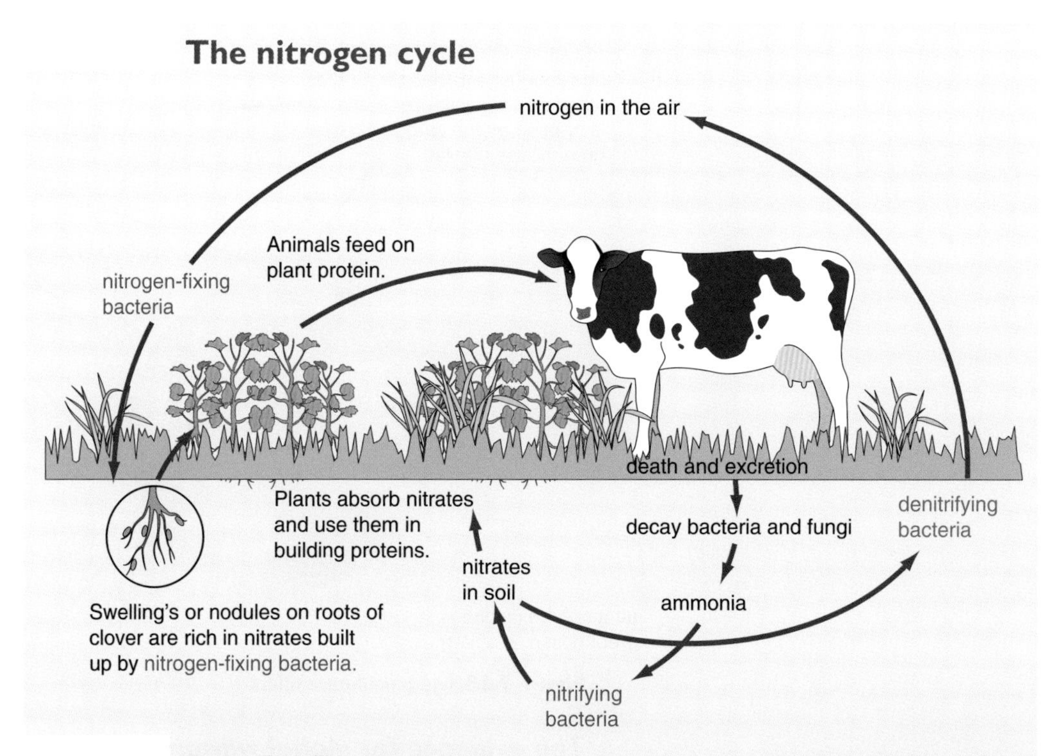

Figure 7.19 The nitrogen cycle

Most of the nitrogen in plants and animals is in the form of amino acids and protein. This is a more complex cycle than the carbon cycle and in understanding the cycle it is worthwhile splitting it into three phases. These three phases are:

* *The build up of nitrogen into amino acids and protein in plants and animals and the eventual breakdown of these compounds into nitrates.* Plants absorb nitrogen as nitrates and use them to make protein. As plants (and animals) are eaten the proteins are eaten, digested and then built up into other proteins in sequence. Eventually the nitrogen is returned to the ground as urine or through the

process of death and decay. Decay or **putrefying bacteria** and **fungi** break down the proteins to release ammonia. A second very important group of bacteria, **nitrifying bacteria**, convert the ammonia or ammonium compounds into nitrates (**nitrification**) and the cycle can continue.

* ***Nitrogen-fixing bacteria*** *are a special group of bacteria that can convert nitrogen gas into nitrates.* These bacteria can be found in the soil or frequently in small swellings (root nodules) in the roots of a particular group of plants called **legumes**. Legumes include peas, beans and clover. The relationship between the legumes and the bacteria is complex, but an important one in which both partners benefit. The bacteria gain carbohydrates from the legumes and they in turn provide a ready source of nitrates for the benefit of the plants. The process of converting nitrogen from the atmosphere into nitrates is called **nitrogen fixation**.
* ***Denitrifying bacteria*** *convert nitrates into atmospheric nitrogen.* This is a wasteful and undesirable process. Denitrifying bacteria are anaerobic and are most commonly found in waterlogged soils. Their effect in well-drained soils is much reduced. The process of converting nitrates into nitrogen is called **denitrification**.

The processes of nitrification and nitrogen fixation can be accelerated by higher temperatures and well-aerated soil.

Root hair cells

Root hair cells are specialised cells in roots that are adapted by having a large surface area for the uptake of nitrates, other minerals and water.

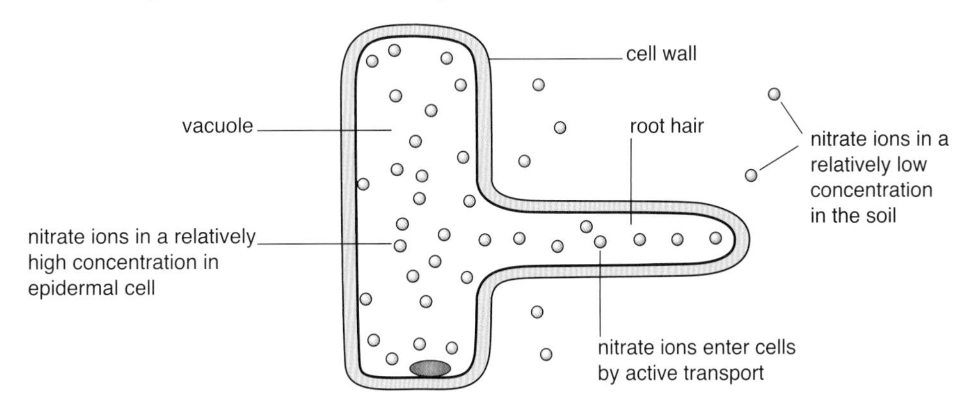

Figure 7.20 Active uptake of nitrates in an epidermal cell

Figure 7.20 shows that there are more nitrate ions inside the root hair cell than outside it. The nitrate ions are taken into the root hair cell by the process of active uptake (transport). This process requires **oxygen** for **aerobic respiration** to produce the energy needed to move the nitrates *against* the **concentration gradient**.

Replacing lost nitrogen – the use of fertilisers

Agricultural land needs to be fertilised on a regular basis. When plants are harvested and animals are taken to the abattoir the nutrients they took from the soil are not replaced.

Calcium (needed for plant cell walls), **magnesium** (needed to make chlorophyll) and **nitrogen** (needed to make amino acids and protein for growth) need to be replaced to maintain plant growth.

The properties of **natural fertiliser** (manure and compost) and **artificial fertiliser** are summarised in the table below.

Natural fertiliser		Artificial fertiliser	
Advantages	Disadvantages	Advantages	Disadvantages
Less soluble than artificial fertiliser so less will be lost by leaching and run-off into waterways	Difficult to store and spread	Easier to store	Soluble so can be easily washed away creating pollution problems
Improves soil quality by adding to the humus content of the soil	Difficult to know the mineral composition exactly	Can be applied in a more controlled manner	Has to be purchased

Addition of nutrients to water from sewage and fertiliser run-off

↓

Leads to increased growth of aquatic plants

↓

When the plants die they are decomposed by bacteria

↓

The bacteria use up the oxygen in the water

↓

Fish and other animals die without oxygen

Figure 7.21 Summary of eutrophication

► Water pollution

Water pollution can have a particularly harmful effect as our rivers, lakes and seas are relatively fragile environments and easily damaged. Every so often we hear about substantial fish kills in local rivers or lakes. This may be due to harmful chemicals being accidentally released from a factory, but many fish kills are a result of sewage or fertiliser run-off draining from farmland that borders the river or lake. The sewage (or slurry) and fertiliser run-off adds to the nitrate level in waterways.

But why does the increased nitrate level kill the fish? The high nitrate levels cause the aquatic plants in the water to grow much faster by providing the nitrogen needed for growth. This may block the light for plants lower in the water. When the plants die they are decomposed by bacteria. The bacteria use up the oxygen in the water and the fish and other animals die due to lack of oxygen. This process is called **eutrophication**.

This type of pollution can be reduced by increasing environmental awareness in farmers, better control of fertiliser use and more secure storage of farmyard manure and slurry.

Acid rain

Waterways (and land) can also be polluted by acid rain. Fossil fuels produce sulfur dioxide (and other gases including nitrogen oxides) and when the sulfur dioxide dissolves in water, acids such as sulfuric acid are produced, which can fall as acid rain. Much of the sulfur dioxide produced in Britain is produced by the burning of fossil fuels in power stations and other large-scale industrial burning.

One of the main problems with acid rain is that it often falls in other countries well away from those causing most of the pollution. This is because the prevailing winds carry the clouds that produce the acid rain for hundreds of miles before the rain actually falls.

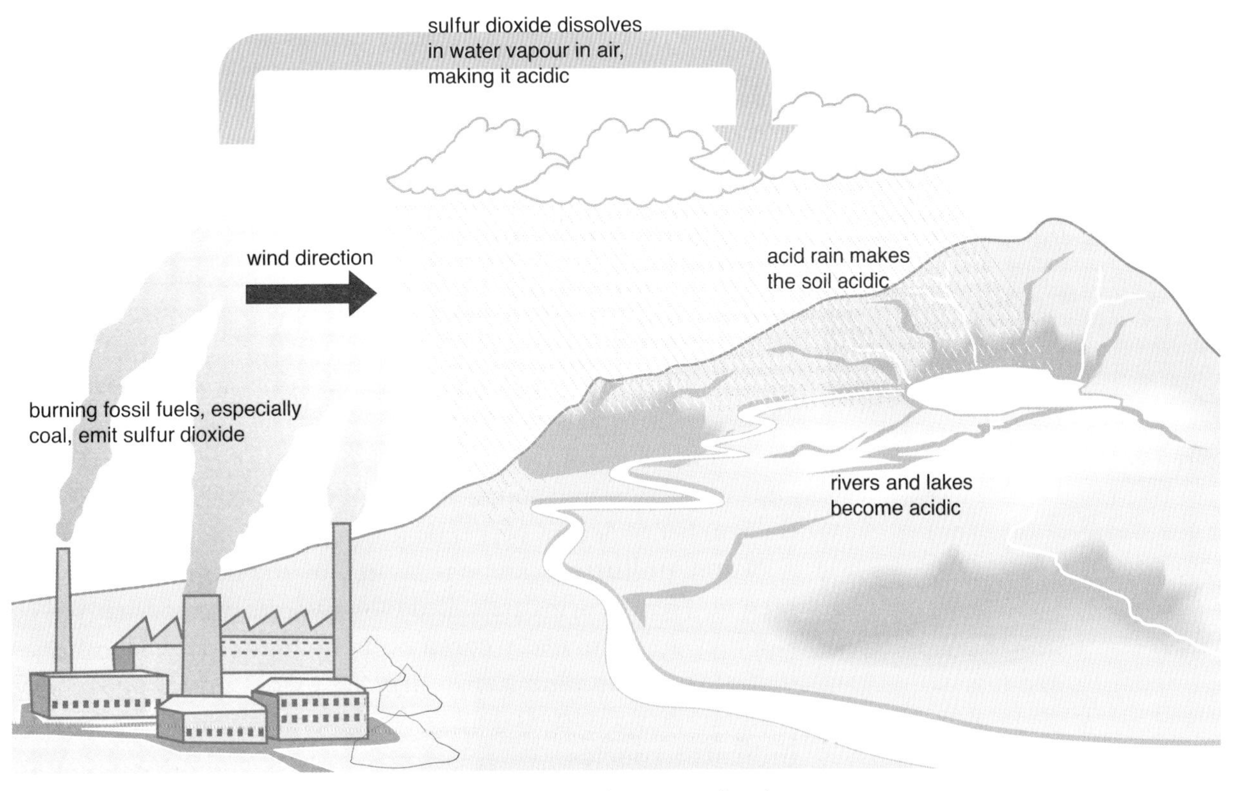

Figure 7.22 How acid rain is produced

But what is the effect of the acid rain on the area on which it falls? Over much of Europe, trees have been destroyed by the effects of acid rain. The acid rain causes the soil to become more acidic and this often means that the trees are no longer living in good growth conditions. The leaves and needles fall off and the trees eventually die.

Rivers and lakes are also badly affected. The acid rain causes the water to become much more acidic, which poisons fish.

Acid rain has become an important international issue because its effects cross international borders as described above. One way of reducing the release of sulfur dioxide and other harmful chemicals from power stations and factories is to use filters that trap these harmful gases. There is also a greater awareness that other sources of generating electricity, such as wind, water, solar and even nuclear power, may be better long-term solutions than fossil fuels.

It has been estimated that Britain has been able to reduce its sulfur dioxide emissions by about 80% over the last 25 years. This has resulted in the recolonisation by native species of many lakes that were seriously damaged by acid rain.

▶ Methods of monitoring change in the environment

It is important that changes in the environment are monitored very closely, whether they are physical changes such as global warming or changes to the numbers or types of animals and plants in an area. An important physical or **abiotic** factor that may be measured is carbon dioxide levels.

Other abiotic factors that can be measured to determine the rate at which global warming is occurring include the extent and size of the polar ice fields, ice density and sea levels.

Figure 7.23 Lichens

The number and distribution of plants and animals can also provide information about environmental change. This type of information is called **biotic** data. **Lichens** are very simple plants that often grow on the roof tiles of houses and also on the bark of trees. Lichens are often yellow or light-grey in appearance.

Most lichens will not grow where there are high levels of **air pollution**. Not surprisingly they are relatively rare in industrial towns and cities but become quite abundant in many rural areas. By monitoring the number of lichens over time it is possible to monitor pollution levels.

Water pollution can be measured by the number of **bloodworms** present. These worms are called bloodworms due to their red colour and they are particularly common in polluted water. They are a good indicator of eutrophication as they are particularly abundant in water with low oxygen levels. Lichens and bloodworms are examples of **indicator species**.

▶ The role of the Government in conserving the environment

The development of a framework for conservation is one of the responsibilities of **Government**. Strategies to encourage conservation include the following:

Reduction of carbon emissions

The reduction of carbon emissions can be encouraged by legislation, for example decisions on energy sources, limits on carbon dioxide emissions, regulations on minimum standards for house insulation or grants for installing solar heating. Education and advertising can help individuals play their part in reducing carbon emissions.

Increasing renewable energy

It is the Government's decision to make overarching decisions that encourage an increase in renewable energy. This includes key planning decisions on nuclear power stations and other forms of energy production such as wind farms. Providing grant aid for household solar power also plays a part. In addition the Government can fund research into the development of different types of renewable energy sources.

Changes in agriculture

The Government is responsible for ensuring that regulations to limit fertiliser application and maintain river quality are in place. They can also encourage sustainable development by providing financial and other support for projects such as growing willow for biofuel.

These areas are not mutually exclusive and there is considerable overlap between them – for example, using most of the renewable energy sources will also help to reduce carbon emissions.

Question

11 Explain why the growing of willow for biofuel is an example of sustainable development.

Figure 7.24 Turbine for generating electricity from tidal energy in Strangford Lough

▶ International co-operation and legislation

The importance of international co-operation in combating pollution was evident in the development of the Kyoto Protocol in 1997. The Kyoto Protocol was a summit that produced a widespread agreement on approaches to reduce the problems associated with climate change. A key agreement reached at the summit was that countries would reduce their pollution levels to the extent that levels produced in 2010 would be no more than the levels produced in 1990. Much of the focus was on carbon dioxide and a pollution credit system was proposed where rich industrialised countries could purchase carbon credits from poorer, less industrialised countries. The effect would be to reduce pollution at a global level with high levels of carbon dioxide pollution in some countries being compensated for by low levels in others.

The difficulties in maintaining widescale international agreements were highlighted by the USA withdrawing its commitment to meeting the agreed Kyoto targets in 2003. In addition, developing countries argue that (Western) developed countries shouldn't be trying to halt the progress of developing countries towards becoming industrialised and developed nations. There have been (and will be in the future) other attempts to agree an international consensus on approaches to climate change. However, the conflict between economic development and conservation in individual countries means that progress will be slow.

Question

12 Suggest how the European Nitrates Directive can reduce eutrophication.

Other international treaties include the European Nitrates Directive. This directive ensures that farmers have adequate storage facilities for farmyard manure and slurry and that some types of fertiliser are not used during the winter months. There is a 'closed season' for the application of farmyard manure and artificial fertilisers. This ensures that the fertiliser is only added at times when plants are growing and it can be used.

▶ Exam questions

1 The bar charts show the results obtained by a class of pupils during a seashore investigation. They sampled the numbers of different seaweeds found in 0.25 m² from the top of the shore (upper shore) to the lower part of the shore and found the average percentage cover for each seaweed.

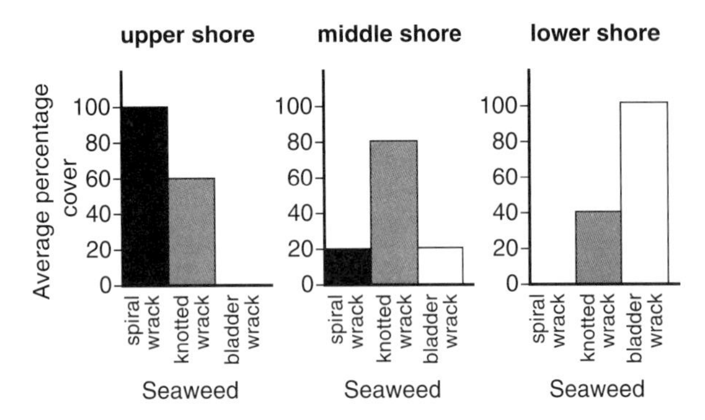

a) i) Describe the trends for the bladderwrack and spiral wrack from the upper to the lower shore. *(1 mark)*

ii) Suggest **one** way that conditions may vary between the upper and lower shore on a seashore, which may have an impact on the survival of living organisms. *(1 mark)*

iii) Seaweeds are a type of algae (plant). Name **one** other physical factor that might affect their growth and suggest how this factor could be measured. *(2 marks)*

b) Describe how the pupils would have carried out this investigation and explain how they would have obtained these results.
In this question you will be assessed on your written communication skills including the use of specialist science terms. *(6 marks)*

c) The table gives the numbers of organisms from a seashore food chain:

large seaweeds → winkles → crabs → gulls

Name of organism	Numbers
Gulls	6
Crabs	10
Winkles	30
Large seaweeds	4

i) Plot a pyramid of numbers for this food chain on graph paper. Label each trophic level with the name of the organism. *(3 marks)*

ii) Suggest **one** problem in obtaining accurate numbers of animals in this food web. *(1 mark)*

d) To obtain the total number of winkles, their numbers were recorded in an area of 0.25 m² and the process repeated ten times along the seashore.
The results were as follows:

	Sample Number									
Number of winkles	1	2	3	4	5	6	7	8	9	10
	3	1	7	4	2	5	1	4	2	1

i) Calculate the average number of winkles per 0.25 m² (show your working). *(1 mark)*

ii) Using this information, calculate the average number of winkles per m². *(1 mark)*

e) In a similar study a month later the winkle population was found to be five per m². Suggest **two** factors which would account for the change in population size. *(2 marks)*

2 The graphs show the summer populations of pied wagtails and swallows in Britain between 1962 and 1965.

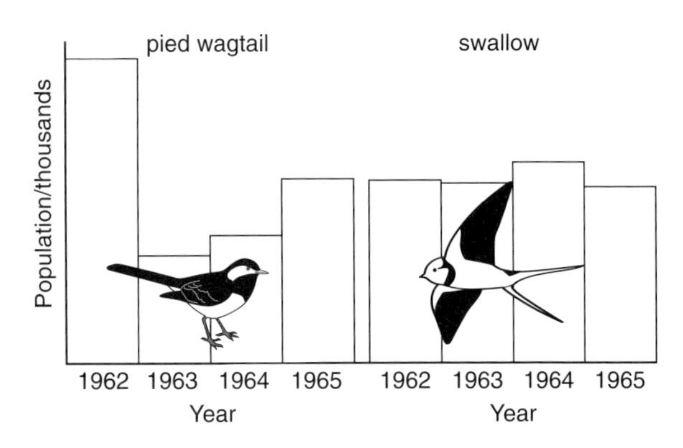

The pied wagtail is resident in Britain 12 months of each year while the swallow is only resident during the summer months.
Compare changes in the populations of these species between 1962 and 1965. Use the data above and your scientific knowledge to explain these changes and how they may be linked.
In this question you will be assessed on your written communication skills including the use of specialist science terms. *(6 marks)*

3 The key shows a system of classification using five kingdoms.

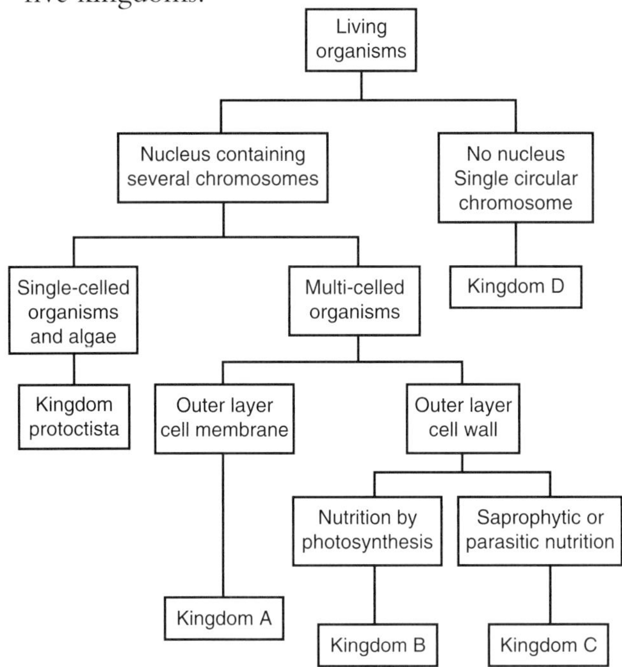

a) Use the key to identify the kingdoms A, B, C and D. *(4 marks)*

b) The smallest group that living organisms can be classified into is a species. What is a species? *(2 marks)*

c) Explain why viruses are difficult to classify. *(2 marks)*

4 The diagram shows a food web for an island in the Arctic.

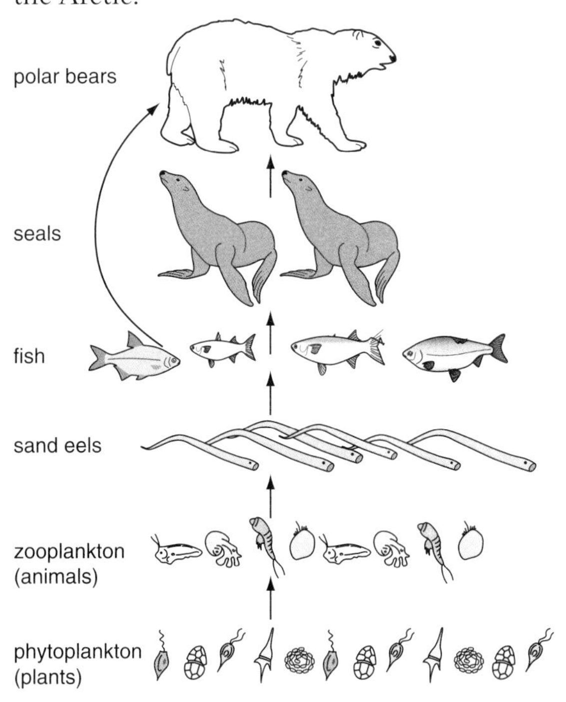

i) Name the energy source for this food web. *(1 mark)*

ii) Use the food web to name:
a primary consumer
a secondary consumer. *(2 marks)*

iii) Use the food web to give a food chain with **only** five types of organisms. *(3 marks)*

iv) Explain why the phytoplankton (plants) are called producers. *(2 marks)*

v) Suggest why the numbers of zooplankton may decrease during winter. *(2 marks)*

vi) On a grid draw a pyramid of numbers for your food chain in (iii). Use the numbers of organisms shown in the food web. Beside each level write the name of the organism. *(5 marks)*

vii) Explain why it is an advantage to the polar bear to have more than one food source. *(1 mark)*

5 The graph below shows changes in the area of Arctic ice.

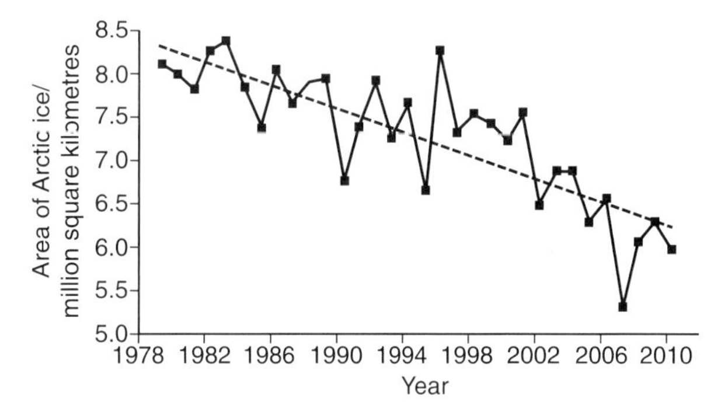

Scientists are researching reasons for these changes. The research suggests that an increase in levels of carbon dioxide leads to global warming. The graph below shows the concentration of carbon dioxide in the atmosphere.

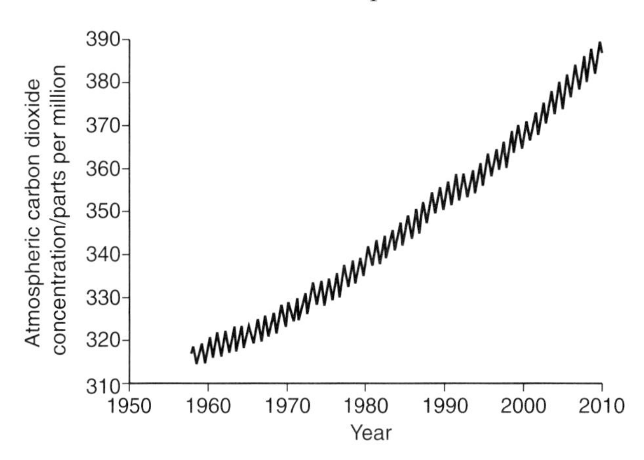

Describe the trends shown in each graph. Explain how these trends provide evidence for the theory that changing levels of carbon dioxide in the atmosphere are linked to the changes in the area of Arctic ice.
In this question you will be assessed on your written communication skills including the use of specialist science terms. *(6 marks)*

6 The diagram shows the nitrogen cycle.
A, B and C are processes carried out by bacteria.

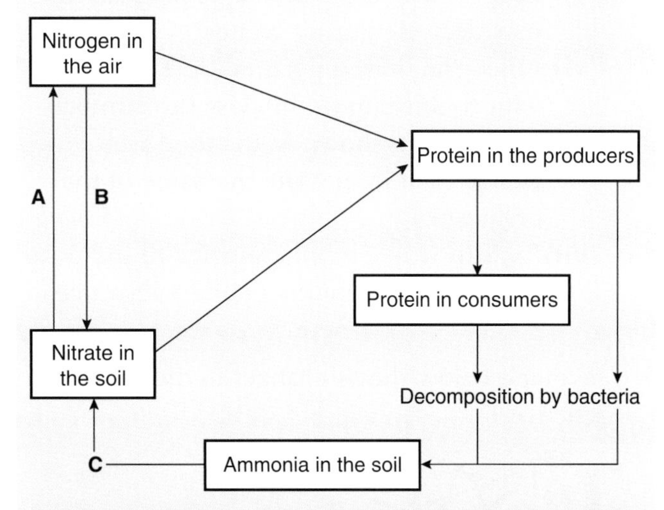

Use the diagram and your knowledge to answer the following questions.

a) Bacteria are decomposers. Name another type of decomposer. *(1 mark)*

b) Name the types of bacteria that carry out the following processes: A, B, C. *(3 marks)*

c) If a farmer is growing a crop, which of these types of bacteria is not helpful? *(1 mark)*

d) Planting clover increases the number of type B bacteria as these bacteria are found in swellings in clover roots. Suggest the benefit to the soil of planting clover. *(1 mark)*

7 a) The following diagram shows the nitrogen cycle.

i) Using the diagram and your knowledge, name the nitrogen bacteria type 1 and type 2. Describe each of their roles in the nitrogen cycle. *(4 marks)*

ii) Explain the effect on plants if nitrogen bacteria type 1 and type 2 were not present in the soil. *(2 marks)*

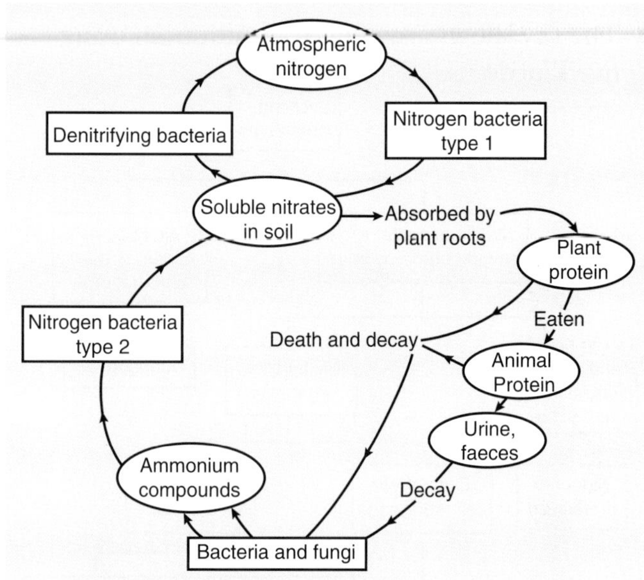

iii) How are plant root cells adapted for absorption of nitrates? *(1 mark)*

b) Water pollution occurs when substances such as fertilisers or raw sewage enter rivers. The graphs show some of the effects of raw sewage on rivers.

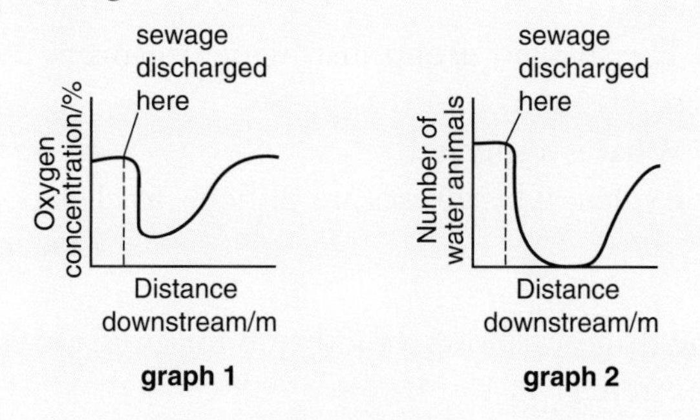

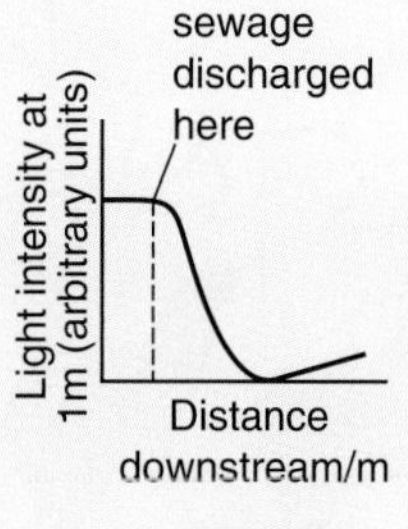

i) Using **graphs 1** and **2** to help you, describe and explain the effect of the raw sewage on the number of river animals. *(3 marks)*

ii) Using **graph 3** to help you, describe and explain what causes the change in light intensity and how it affects the water plants near the sewage discharge. *(3 marks)*

8 The shaded areas on the maps show the distribution of lakes in Norway where fish stocks have been killed by acid rain.

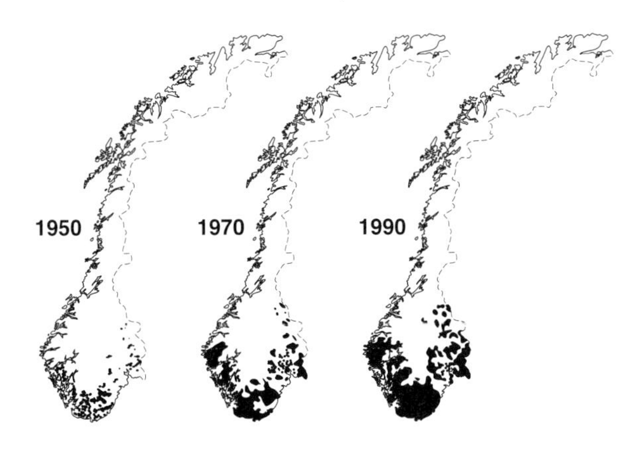

a) Describe the trend shown in the maps. *(1 mark)*

b) Give **one** other biological effect of acid rain. *(1 mark)*

Sulfur dioxide, produced when fossil fuels are burned in other parts of Europe, is an important gas involved in the formation of acid rain.

c) Explain how sulfur dioxide forms acid rain which reaches lakes in Norway. *(2 marks)*

Since 2000 the number of lakes affected has reduced. Surveys of fish populations are one way of monitoring such environmental change.

d) Give **one** other example of biotic data which can help monitor pollution. *(1 mark)*

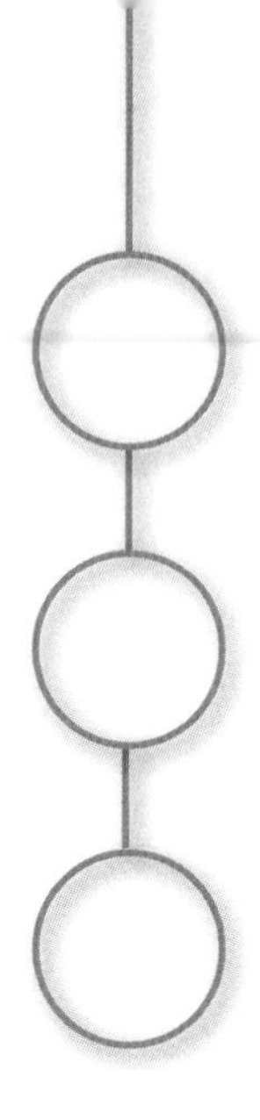

8 Osmosis and Plant Transport

The movement of substances into and out of cells

Cell membranes allow some substances to pass through but prevent the movement of others – they are **selectively permeable**. Plant cells are also surrounded by a cell wall. The cell wall is totally permeable and has no role in controlling what enters or leaves cells.

Diffusion, the random movement of a substance from where it is in high concentration to where the concentration is lower, is an important process in the transport of substances through cell membranes. We have already met diffusion in Chapter 1.

Osmosis

Osmosis is a special type of diffusion involving the movement of water molecules through a selectively permeable membrane. If pure water and a sugar solution are separated by a selectively permeable membrane (e.g. dialysis or Visking tubing), then osmosis will occur. The water will move from where it is in a higher concentration (e.g. pure water) to where it is in a lower concentration (sugar solution). Another way of describing this is that the water moves from the weaker to the stronger solution.

Osmosis can be defined as the movement of water from an area of greater concentration of water to an area of lower concentration of water through a selectively permeable membrane. Figure 8.1 shows that in osmosis, water molecules move but other larger molecules such as sugar cannot fit through the selectively permeable membrane.

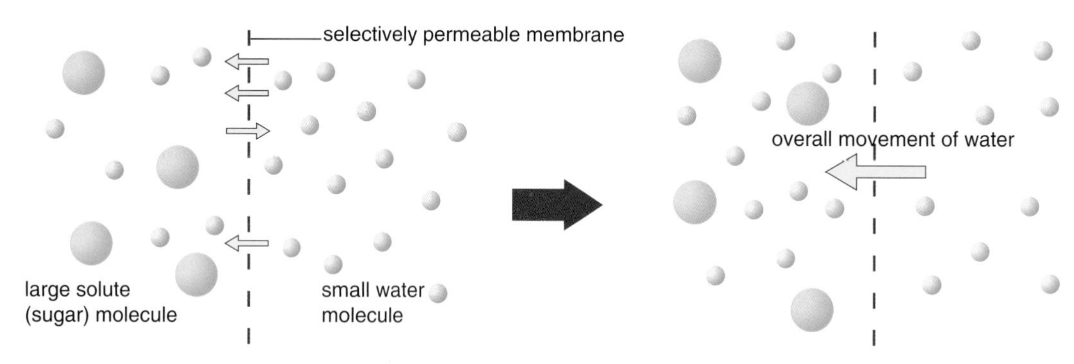

A concentration gradient exists across a selectively permeable membrane. The water molecules can move in any direction across the selectively permeable membrane. There is a higher concentration of water on the right side of the membrane so there will be a net movement of water to the left.

The water molecules have moved through the selectively permeable membrane until there is the same concentration on each side.

Figure 8.1 Osmosis

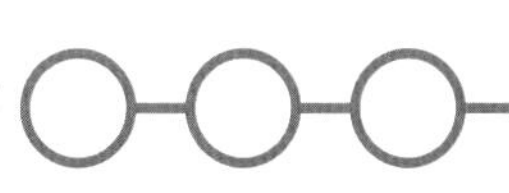

Question

1 a) If a potato cylinder is placed into pure water it will gain mass. Explain why.
b) What would happen if a potato cylinder was placed into a strong sugar solution? Explain your answer.
c) Suggest why a potato cylinder placed in a weak sugar solution had the same mass when reweighed after a period of time.
d) Instead of measuring mass you could measure the length of the potato cylinder. Suggest two reasons why it is more accurate to measure mass.

Osmosis is very important in the movement of water into and through the cells of plants and animals.

Osmosis in animal cells

If a red blood cell is placed in pure water (or a very weak solution), water will enter the cell by osmosis. This happens because a concentration gradient exists and water will move down the gradient into the cell. In this example so much water would enter the red blood cell by osmosis that it would swell up and eventually burst (**lysis**). Conversely, if a red blood cell (or any animal cell) loses too much water by osmosis then the cell shrivels up.

Clearly, these changes do not happen to cells in a healthy body. The reason our cells do not burst or shrivel up is because the concentration of the blood is carefully controlled to ensure that large volumes of water do not enter or leave the blood, or other cells, by osmosis.

Osmosis in plant cells

Water also moves into and out of plant cells depending on the concentration of the solution surrounding the cell. When water moves into a plant cell the vacuole increases in size, pushing the cell membrane against the cell wall. In a plant cell the cell wall prevents too much water from entering and the cell bursting as would happen in animal cells. The force of the membrane pushing against the cell wall makes the cell firm or turgid. **Turgor** gives the cell support and in non-woody plants is essential in keeping the plant upright. The importance of turgor in providing support can be seen when there is a shortage of water. When plant cells do not receive enough water they cannot remain turgid and wilting occurs. Cells that are not turgid are described as being **flaccid**.

If a plant cell loses too much water by osmosis a condition called **plasmolysis** occurs. During plasmolysis so much water leaves the cell that the cell contents shrink, pulling the cell membrane away from the cell wall. Although loss of turgor and wilting is a common occurrence in many plants, plasmolysis is much less likely in healthy plants. This is just as well, as plasmolysed cells are unlikely to survive!

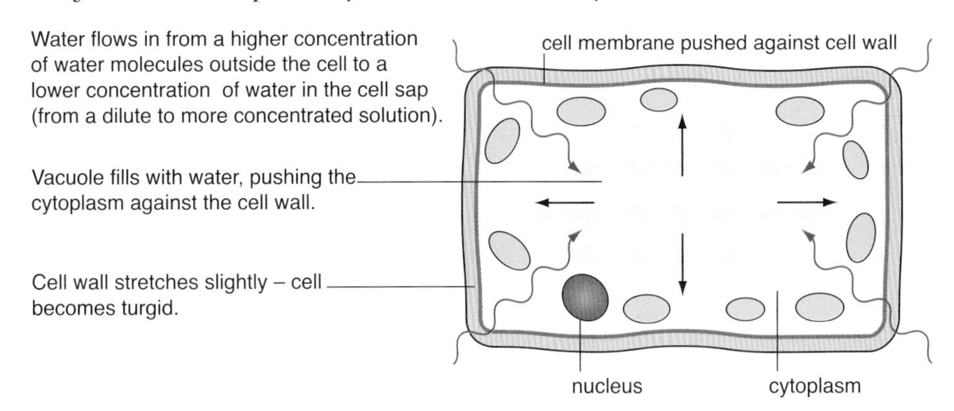

Figure 8.2 This plant cell is surrounded by pure water

Question

2 Explain why plant cells do not burst when placed in water.

Search

▶ osmosis and experiments

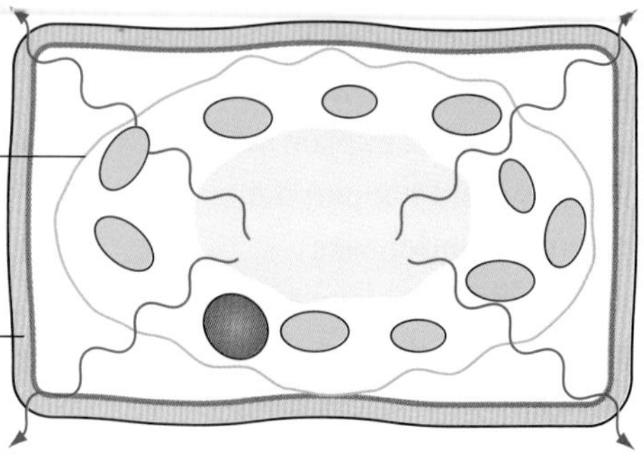

Figure 8.3 This plant cell is surrounded by a solution with a lower concentration of water molecules

You should be able to use a microscope to observe turgid and plasmolysed cells. Revise how to use a microscope in Chapter 1.

▶ Transpiration

Much of the water that enters the leaves evaporates into the atmosphere. This loss of water by evaporation is called **transpiration**. The evaporation of water takes place mainly from the spongy mesophyll cells in plant leaves, through the air spaces and out of the stomata (small pores). This continuous movement of water through a plant (the transpiration stream) is very important for three reasons:

1 The supply of water to the leaves as a raw material for **photosynthesis**.
2 The **transport of minerals** in the water through the root and up the xylem to the leaves and other parts of the plant.
3 As water passes through the plant it enters cells by osmosis to provide **support**.

Summary: The previous paragraph describes the four main functions of water in plants. These are:

1 Providing water for transpiration.
2 Providing water for photosynthesis.
3 Providing water for the transport of minerals.
4 Support.

Plants often need to reduce water loss by transpiration. They do this by closing the stomata that are most common on the underside of leaves. The stomata are necessary to allow gases to enter and leave the leaf but they can be closed on occasions when it is important to conserve water.

Measuring water uptake using a potometer

The **bubble potometer** is a piece of apparatus designed to measure water uptake in a leafy shoot. As water evaporates from the leaves of the cut shoot, the shoot sucks water up through the potometer. The distance an air bubble moves in a period of time can be used to calculate the rate of water uptake.

The reservoir (or syringe) allows the apparatus to be reset so that replicate results can be recorded or the water uptake measured in different environmental conditions. Air leaks will hinder the uptake of water into the plant so it is important that the potometer apparatus is properly sealed, particularly at the junction between the shoot and the neck of the potometer. To prevent the development of unwanted air bubbles in the water column entering the plant, it is necessary to assemble the apparatus under water initially.

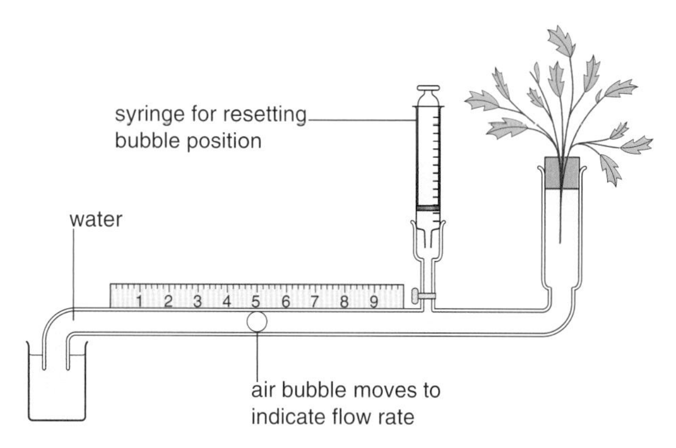

Figure 8.4 The potometer

The potometer can be used to measure how environmental conditions such as **wind speed**, **temperature** or **humidity** affect the rate of water uptake. Conditions that increase the rate of evaporation (and transpiration) such as higher temperatures or higher wind speed (through using a fan) will increase the rate of water uptake. Higher levels of humidity (created through covering the shoot with a polythene bag) will reduce evaporation and transpiration and therefore slow the rate of water uptake by the shoot.

Obviously when changing a particular environmental factor (e.g. wind speed) to show how it affects water uptake, it is necessary to keep other environmental factors constant (e.g. temperature, humidity, light, etc.). This ensures that the results are **valid**. Replicated results will help ensure that the results are **reliable**.

Leaf surface area, while not an environmental factor, will also affect transpiration rates. Larger (or more) leaves will have more stomata through which water can evaporate and diffuse. You should also be able to plan an experiment to investigate the effect of leaf surface area on transpiration rate.

Note: The potometer can accurately measure the volume of water taken up by the shoot but it cannot give an absolute value for transpiration itself. Transpiration is the loss of water by evaporation from the leaves and it is impossible to calculate how much of the water taken up by the plant is actually transpired through the leaf surface. Some of the water will be used in photosynthesis and in providing support through turgor, so the volume transpired will inevitably be less than the volume taken into the shoot. However, it is an excellent method for comparing rates in different conditions.

Question

3 a) Describe how you could introduce the 'bubble' into the water column in a bubble potometer.

b) Describe how you could use a bubble potometer to compare the rates of water uptake in still and windy conditions.

Transpiration rates can also be measured using a **weighing method** – weighing how much water a plant loses over a period of time.

Question

4 Describe how you would use the weighing method to compare the loss of water from a pot plant in humid and in non-humid conditions. Your answer should identify the variables you will keep constant and any other precautions you will make to ensure you get valid results.

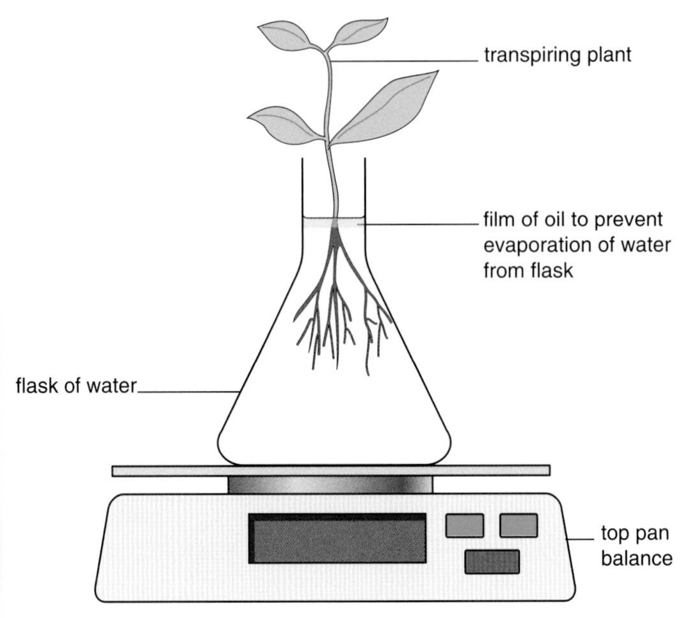

Figure 8.5 The weight potometer

Exam questions

1 An investigation into osmosis in potato cores was set up as shown in the diagram below.

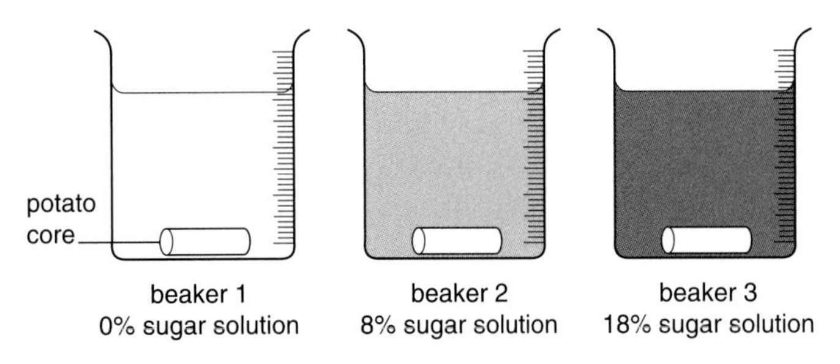

a) Name **two** factors that should be controlled during the investigation in order to make it a fair test. *(2 marks)*

b) How would you know if water had moved out of the potato cores during the investigation? *(1 mark)*

c) In which beaker would you expect most water to move out of the potato cores during the investigation? *(1 mark)*

2 The diagram shows a sugar solution separated from water by a selectively permeable membrane.

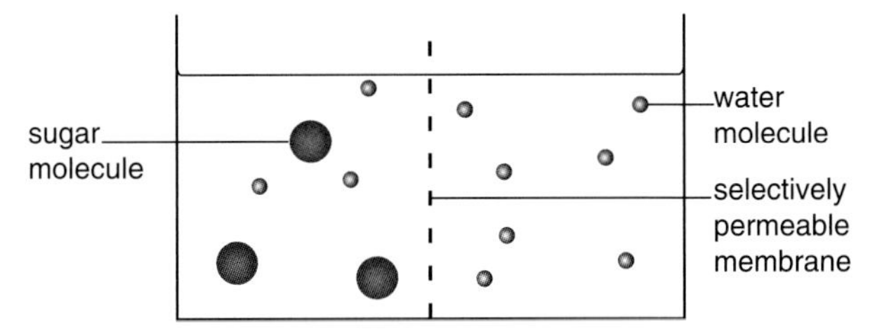

a) Describe and explain what you would expect to happen over a period of time. *(3 marks)*

b) Some pupils carried out an investigation into osmosis in potatoes.
The pupils cut identical lengths of potato, each **50 mm**, at the start of the experiment.
They placed three into water and another three into sugar solution.
They left them for several hours, removed them, dried the potato lengths and remeasured them.

The table shows their results at the end of the experiment.

Length of potato in water/mm	Change in length/ mm	Length of potato in sugar solution/mm	Change in length/mm
51	+1	49	−1
55	+5	44	−6
53	+3	45	−5
	Average change in length = 3 mm		Average change in length = X mm

i) Calculate the average change in length for the potatoes in the sugar solution (**X**). *(1 mark)*

ii) Explain the results for the potatoes in the sugar solution. *(2 marks)*

iii) Suggest **one** factor that the pupils should have kept the same during their experiment. *(1 mark)*

iv) The diagram shows a cell from the potato at the beginning of the experiment.

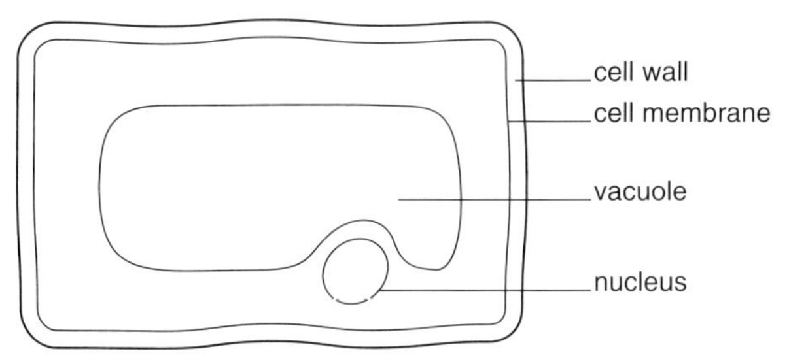

Draw how the same cell would look after it had been in the sugar solution for several hours. *(3 marks)*

3 A pupil carried out the following investigation to find a sugar solution which would prevent potato chips gaining or losing mass.
Potato chips were weighed and placed in test tubes containing a range of sugar solutions.

* After 24 hours the chips were removed, dried on a paper towel and reweighed.
* The experiment was repeated three times for each concentration of solution.

The results are shown in the table.

Concentration of sugar solution in M (Moles)	Initial mass/g	Final mass/g	Change in mass/g (+/−)	% change in mass (+/−)	Average % change in mass (+/−)
0	2.5	3.3	+0.8		
	3.3	4.0	+0.7	21	
	2.0	2.5	+0.5	25	
0.50	3.5	3.4	−0.1	−2.8	−4
	3.1	3.0	−0.1	−3.2	
	1.4	1.3	−0.1	−7.1	
1.00	3.4	2.5	−0.9	−26.5	−33
	3.2	2.0	−1.2	−37.5	
	1.4	0.9	−0.5	−35.7	

a) Calculate:

i) the percentage change in mass for the first chip in 0 molar sugar solution.
Show your working. *(2 marks)*

ii) the average change in mass of the three chips in 0 molar sugar solution to the nearest whole number. *(1 mark)*

b) Why is it necessary to calculate the percentage change in mass for each potato chip? *(1 mark)*

c) Draw a graph of these results, showing the average % change in mass against the concentration of sugar solution. *(4 marks)*

d) i) Describe the trend shown by the results. *(1 mark)*

ii) Use the graph to estimate the concentration of the sugar solution which would prevent potato chips losing or gaining weight. *(1 mark)*

e) Suggest **two** ways in which this investigation should be improved. *(2 marks)*
The diagram shows cells from the potato chips in the strong sugar solution.

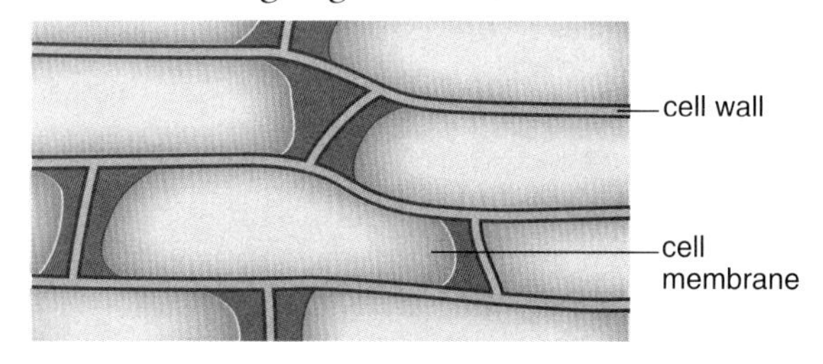

f) Describe and explain in detail, the appearance of these cells.
In this question, you will be assessed on your written communication skills, including the use of specialist science terms. *(6 marks)*

4 The diagram shows part of a leaf and the pathway of water vapour through a leaf.

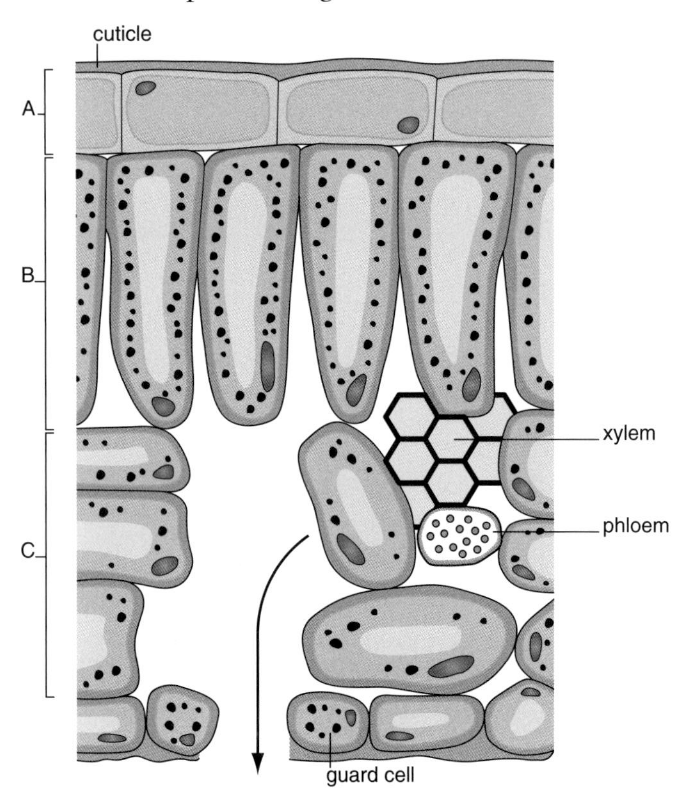

a) Name parts A, B and C. *(3 marks)*

b) Look at the diagram. Explain the movement of water along the arrow. *(3 marks)*

c) Explain how the cuticle and guard cells help reduce the loss of water from the leaf. *(3 marks)*

5 a) The diagram shows apparatus which is used to measure the rate of water uptake in plants.

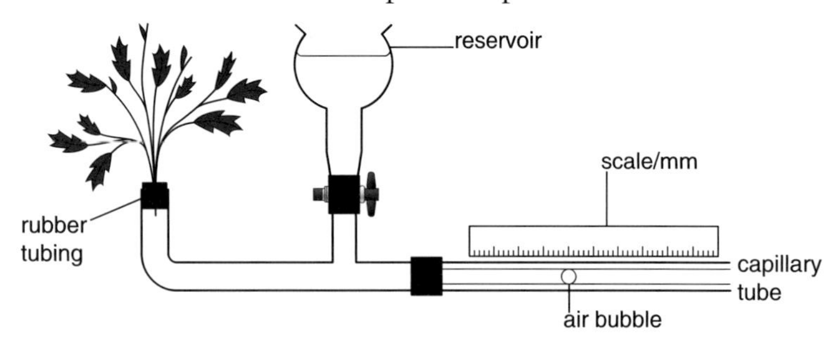

i) Give **one** precaution you should take when setting up this apparatus. *(1 mark)*

ii) Describe and explain how you could use this apparatus to compare the rate of water uptake for a plant shoot at two different temperatures (20 °C and 30 °C).
In this question, you will be assessed on your written communication skills including the use of specialist science terms. *(6 marks)*

b) The table gives one set of results obtained using this apparatus.

Temperature/°C	Distance/mm moved by the bubble after 10 minutes
20	6
30	12

The average rate of movement of the bubble at 20 °C is 0.6 mm per minute.

i) Calculate the average rate of movement of the bubble at 30 °C.
Remember to include units in your answer. *(2 marks)*

ii) Plants grown in warmer conditions wilt quickly and need to be watered more frequently than those grown in normal room temperatures.
Use the evidence from your calculations to account for this difference. *(3 marks)*

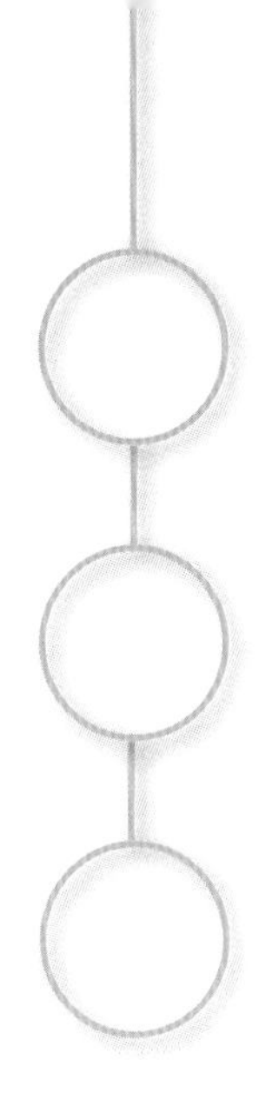

Chromosomes, Genes and DNA

Most living cells contain a **nucleus** (control centre). The nucleus is the control centre because it contains **chromosomes** that are subdivided into smaller sections called **genes**. There are hundreds of genes in each chromosome. Figure 9.1 shows a single chromosome. Normally they occur in **functional pairs** (except sex cells) – we will see why in the next chapter.

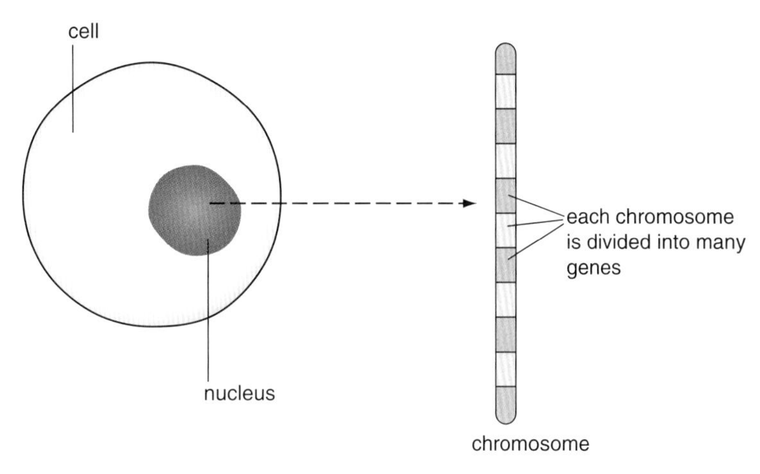

Figure 9.1 The nucleus, chromosomes and genes

It is the genes in our bodies that control characteristics such as eye and hair colour – the features that make us what we are. Inside genes and chromosomes there is a very important molecule that gives them their properties. This molecule is **DNA** (deoxyribonucleic acid). In effect genes are short lengths of DNA that code for a particular protein or characteristic.

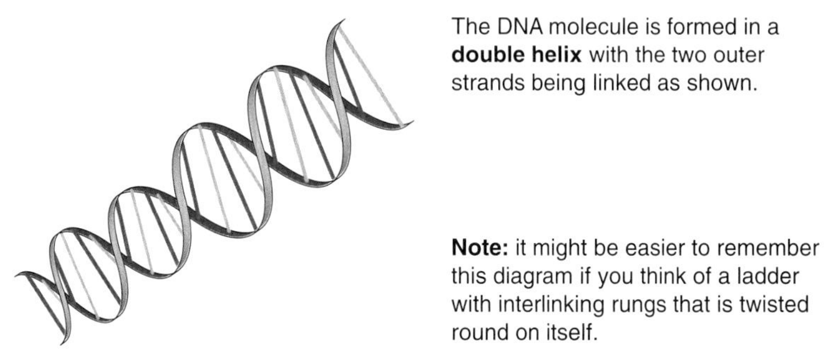

Figure 9.2 A section of DNA double helix

The structure of DNA

DNA consists of three sub-units that are regularly repeated throughout the length of the molecule. These sub-units are **deoxyribose sugar**, **phosphate** and **bases**. There are four different types of base:

adenine, **guanine**, **cytosine** and **thymine**. In the double helix the rungs of the 'ladder' are the bases and the sides are alternating units of deoxyribose sugar and phosphate.

Each repeating unit of DNA, consisting of a phosphate, a sugar and a base, is called a **nucleotide**. Figure 9.3 shows that bases link the two sides of the molecule together in such a way that adenine only combines with thymine and guanine only combines with cytosine. This arrangement is known as **base pairing**. The arrangement of the bases along the length of the DNA is what determines how a gene works.

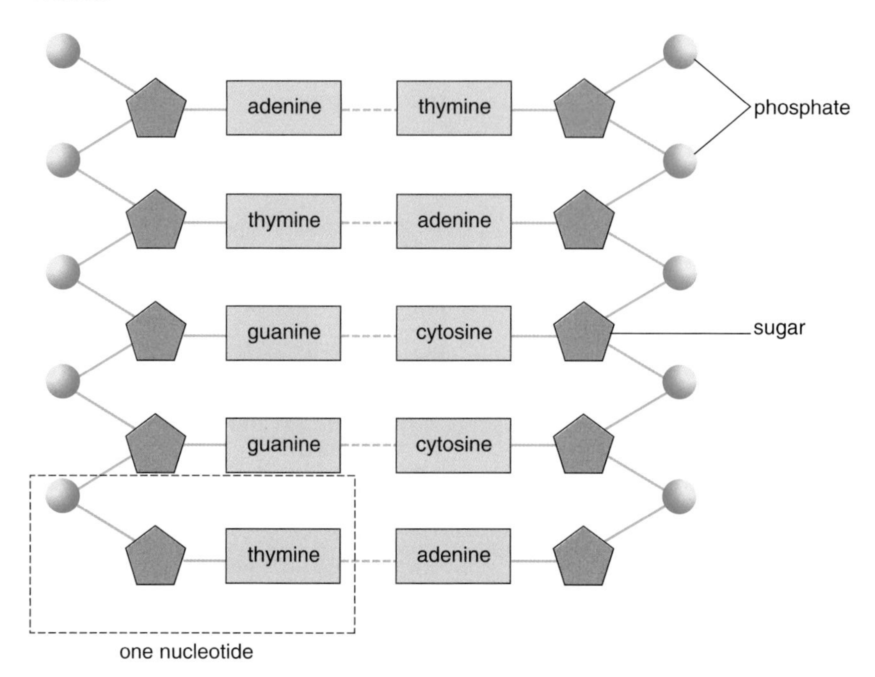

Figure 9.3 Base pairing in DNA

If we map the arrangement of bases along each individual's chromosomes we will find that while there will be similarities among different individuals, no two people have the same sequence of bases along the entire length of all their chromosomes (except for identical twins!).

How does DNA work?

The DNA works by providing a code to allow the cell to make the proteins that it needs. The DNA determines which proteins, and in particular which enzymes, are made. Enzymes are extremely important proteins that control the cell's reactions. Therefore, by controlling the enzymes, the DNA controls the development of the cell and in turn the entire organism.

The bases along one side of the DNA molecule – the coding strand – form the genetic code. Each sequence of three bases (a triplet) along this coding strand codes for a particular amino acid – the building blocks of proteins. The sequence of three bases that codes for an amino acid is called a **base triplet**. As a protein consists of many amino acids linked together it is important that the correct base triplets are arranged in the correct sequence along the coding strand. Figure 9.4 shows how base triplets code for particular amino acids. In Figure 9.4 the first and fourth base triplets have the same code and this means that the first and fourth amino acids are also the same. The model only shows a small section of a gene and a small section of the protein that it produces codes for.

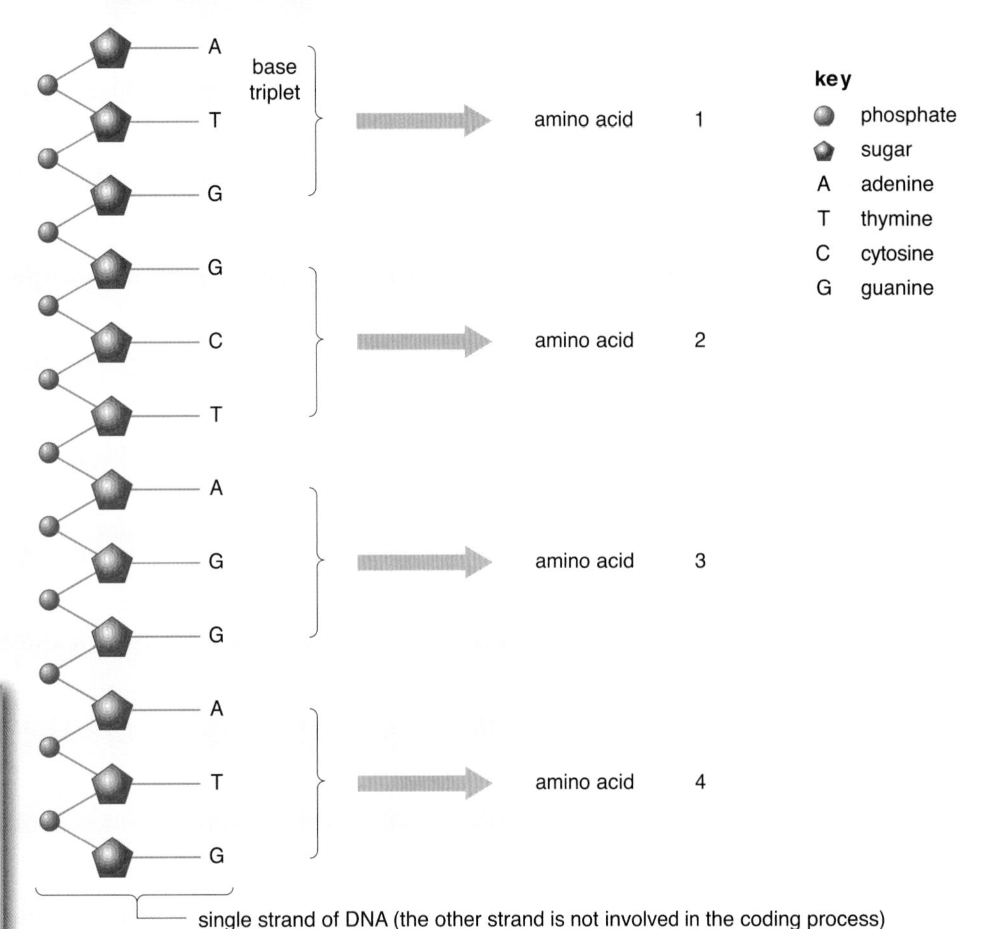

Figure 9.4 How base triplets code for amino acids

Question

1 Use Figure 9.4 to work out the length of DNA required (in number of bases on the coding strand) to code for a protein consisting of 177 amino acids.

Building the theory – working out the structure of DNA

As you can see from the earlier figures, DNA is a very complex molecule. A lot of expertise and hard work was required in working out its structure. In the early 1950s scientists had worked out that DNA was the molecule that determined how organisms developed – they just didn't know its structure!

In 1950 **Erwin Chargaff** discovered that although the arrangement of bases in DNA varied, there was always an equal amount of adenine and thymine. Similarly there was always an equal amount of guanine and cytosine.

Question

2 Explain Chargaff's observation.

The next part of the DNA jigsaw was put in place by **Rosalind Franklin** and **Maurice Wilkins**, who were research scientists working at King's College, London. They used a process called **X-ray diffraction (crystallography)**. In X-ray diffraction, beams of X-rays are fired into molecules of DNA and the ways in which the DNA scatters the X-rays provides information about its three-dimensional structure. Franklin and Wilkins were able to work out the overall shape of DNA but they were not able to confirm exactly how the sub-units were linked together.

The last part of the puzzle was solved by **James Watson** and **Francis Crick** at Cambridge University in 1953. They were able to build on the work of the previous scientists to deduce how the bases were arranged and also to conclude that the DNA molecule is arranged as a double helix. They did this through the process of **modelling**.

Figure 9.5 Watson and Crick with their model of DNA

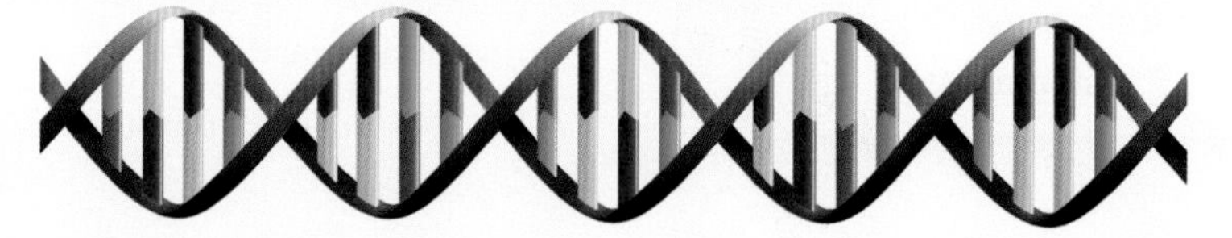

Figure 9.6 Model of DNA

The discovery of the structure of DNA is typical of many scientific discoveries in that theories are often built up in stages with many scientists laying the foundations before the final details are worked out. Another common characteristic of scientific discoveries can be seen in the working out of DNA's structure – the scientists who put the finishing touches to a theory get most of the credit! We have met the process of peer review in validating new scientific knowledge in Unit 1. Peer review was also an important part of the process in confirming the accuracy of each step in working out the structure of DNA.

Exam questions

1 The diagram shows one strand of a DNA molecule.

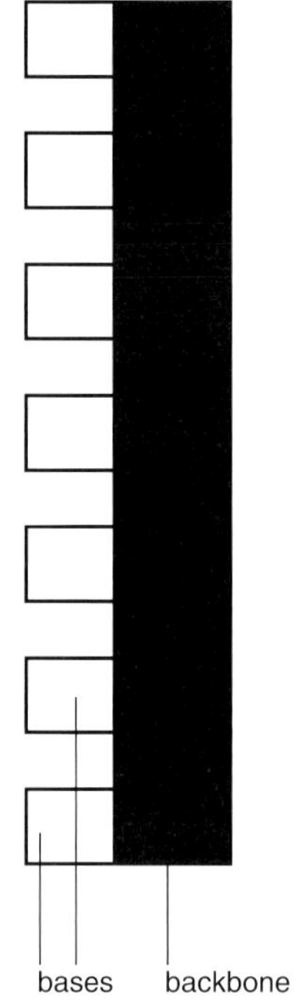

a) Copy and complete the diagram by drawing the missing strand. *(1 mark)*

b) Name the **two** chemicals that make up the backbone. *(2 marks)*

c) Describe the three-dimensional shape of DNA. *(1 mark)*

2 The diagram below shows the 3D structure of DNA, which was discovered by Watson and Crick.

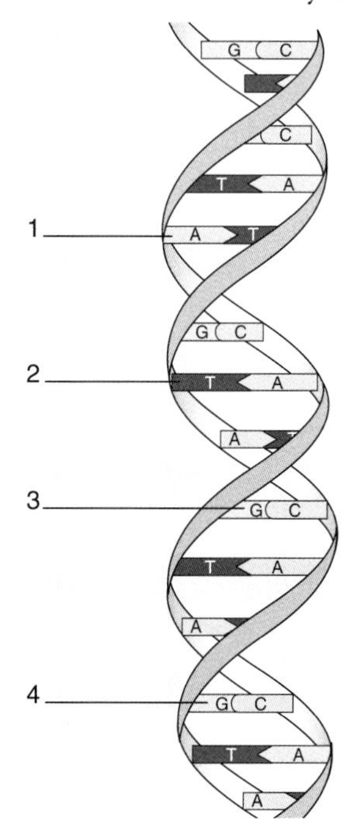

a) Name **two** other scientists who made a discovery about the 3D structure of DNA. *(2 marks)*

b) If there were 18 of these structures along a coding strand of DNA how many amino acids would they code for? *(1 mark)*

3 a) The diagram shows how DNA codes for amino acids.

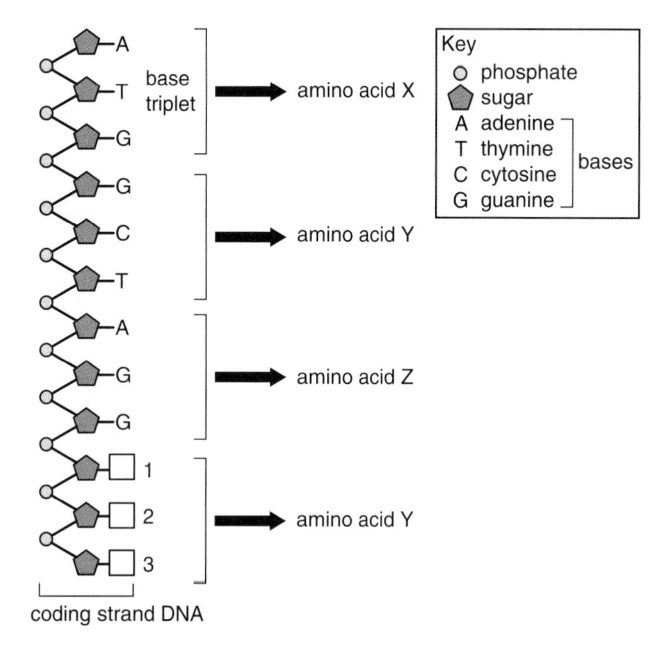

i) Give the names of the missing bases 1, 2 and 3. *(1 mark)*

ii) Use the diagram and your knowledge to calculate how many bases would be required to produce 30 amino acids. *(1 mark)*

b) Franklin and Wilkins showed that the arrangement of DNA resembled a ladder structure although they did not know how the pieces were linked.

i) Describe **two** further advances that Watson and Crick added to our understanding of the structure of DNA and name the process they used. *(3 marks)*

ii) Name **one** other scientist who added to our knowledge of DNA and describe his/her contribution. *(2 marks)*

4 a) Below is a partially completed diagram of the structure of DNA.

i) Copy and complete the diagram by drawing in the missing bases. *(1 mark)*

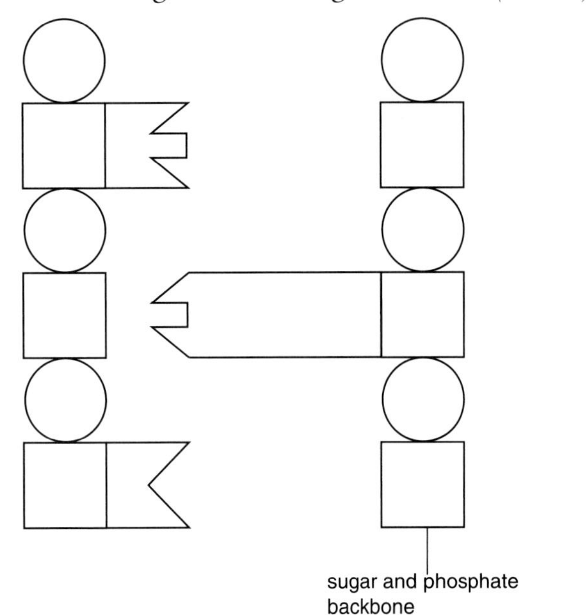

ii) How many different bases does your completed diagram show? *(1 mark)*

iii) State **one** fact that Watson and Crick discovered about these bases. *(1 mark)*

iv) Name **one** other scientist who researched the bases in DNA. *(1 mark)*

b) State how the **base triplet hypothesis** explains the link between the DNA code and the formation of proteins. *(3 marks)*

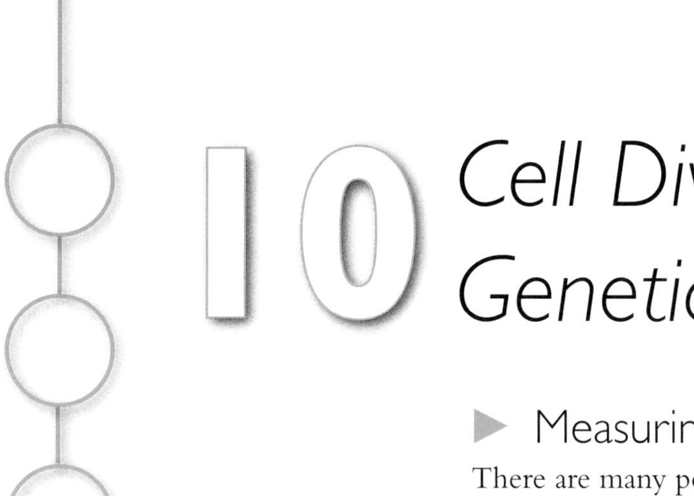

Cell Division and Genetics

Measuring growth

There are many possible methods that could be used to measure growth in living organisms. In humans, two obvious measurements are **height** and **mass**. People attending hospital sometimes have their height and weight (mass) measured – this gives a good indication of how their development compares with average rates.

Question

1 Jane is 20 years old and has had her height and mass measured a number of times over a year-long period. Suggest why her mass showed greater variability than her height.

Other measurements of growth

Method	Advantages	Disadvantages
Cell length	Easy to measure under a microscope	Often irregular shape No reference to number of cells
Number of cells	Good indication of overall size	Organisms may contain millions of cells – difficult to estimate or count them all Doesn't take account of size of cells
Dry mass*	Accurate indication of growth	Time consuming Organism or part (e.g. leaf) is killed during process

Dry mass* is the mass of the organism after all the water content is removed. It is found by drying in an incubator until there is a constant mass.

Questions

2 a) Suggest why the organisms need to be dried until there is a constant mass.

b) Explain why dry mass is likely to be a more accurate measurement of plant growth than mass without drying (fresh weight).

Mitosis

Most living organisms **grow** by increasing their cell number. Cells double in number by splitting in half. It is important when cells divide during growth that the two new (daughter) cells end up with exactly the same genetic makeup as each other and the parent cell – they are **clones** of the parent cell. This means that every cell in the growing organism has the same number and kind of genes and chromosomes, and that these are the same as they were in its very first cell, the zygote. This type of cell division is called **mitosis**.

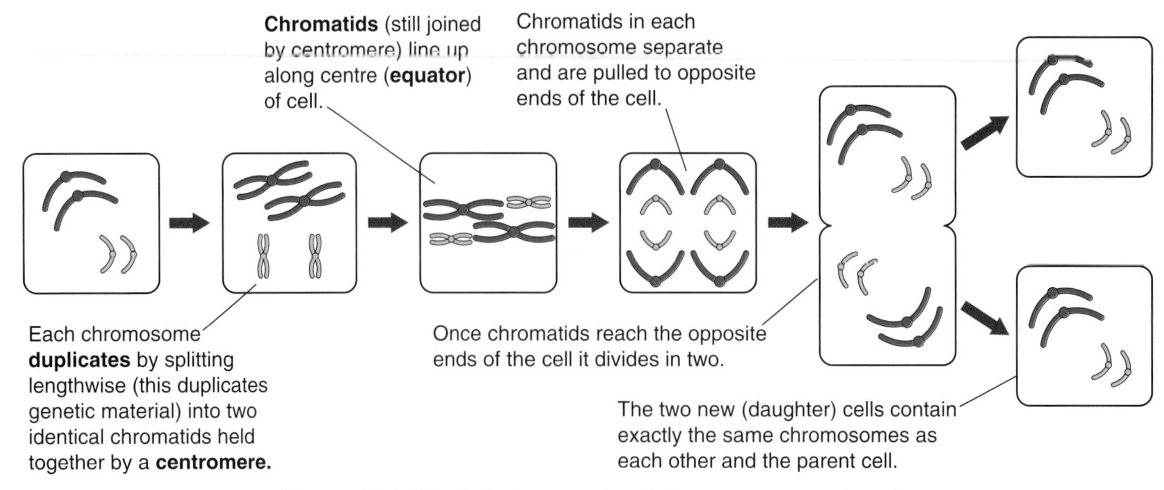

Figure 10.1 Mitosis (only two pairs of chromosomes are shown)

Figure 10.1 shows that the two daughter cells contain exactly the same chromosomes as each other and also as the parent cell from which they were produced.

Mitosis is a type of cell division used in **growth**, to **replace** worn-out cells and to **repair** damaged tissue.

plant with desirable characteristics
sterile scalpel
stem
explants
explant
nutrient jelly
each explant grows into a small plant

Figure 10.2 Tissue culture

Question

3 Use the information opposite to draw a table highlighting the main advantages and disadvantages of plant cloning.

Asexual reproduction in plants

Asexual reproduction, common in plants, produces genetically identical offspring. This is because gametes are not involved and the cells from the adult simply reproduce identical copies of themselves by mitosis to form a new individual. Examples in nature include daffodils forming daughter bulbs and strawberry runners producing new plants. Because the new plants are genetically identical to their parent, they are referred to as clones and the process is called **cloning**. A big advantage with asexual reproduction is that only one parent is needed.

Asexual reproduction or cloning occurs in nature, but we can also deliberately clone other living organisms to make sure that they develop with the characteristics we want. The use of cuttings in plants is a form of cloning that has been taking place for many years. Advantages of plant cloning include the fact that only plants with desirable qualities are produced, i.e. you know what you will get. It is also a quick process of producing lots of plants, and many cuttings can be taken from the same plant.

There are disadvantages with cloning as well. The lack of variation means that all the plants may be equally susceptible to disease or have other weaknesses.

Tissue culture is a particular type of cloning in plants that can be used to produce a very large number of identical plants. It can be carried out in the laboratory all year round, and if carried out in sterile conditions can produce disease-free varieties. Tissue culture is a good way of conserving very rare plants.

Cancer

Cancer is a term used to describe a range of diseases caused by **uncontrolled cell division**. When someone has cancer, the cells in the part of the body affected divide uncontrollably to form a growth or tumour.

Causes of cancer

Cancer can develop for no obvious reason but there are many environmental factors that increase the likelihood of cancer occurring. These include:

UV radiation
UV radiation from the Sun or sun beds can cause **skin cancer**. The rate of skin cancer in Britain is rapidly increasing but the possibility of getting it can be reduced by reducing the time spent in strong sun, covering up or using sun lotion.

Chemicals in cigarette smoke
Cigarette smoke contains a number of cancer-causing chemicals collectively known as **tar**. Tar is a very important cause of **lung cancer**.

Viruses
Some cancers can be caused by viruses, including the **human papilloma virus (HPV)**. There are a number of types of HPV virus and they are linked to various human cancers including **cervical cancer**.

Question

4 John is going to Majorca for a 2-week holiday. Describe three things he can do when there to reduce his chances of getting skin cancer.

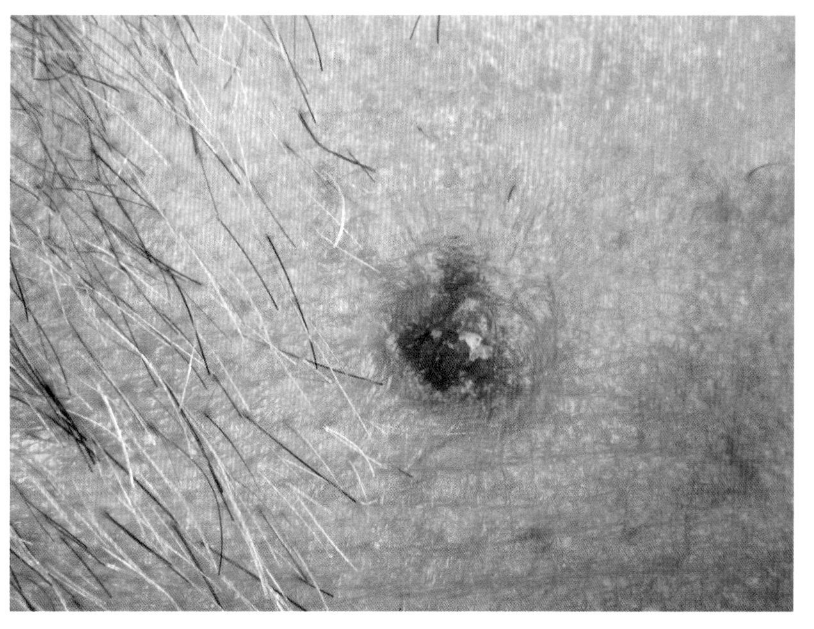

Figure 10.3 Skin cancer

Types of tumours (cancer)

There are two types of tumour:

* **Benign** tumours remain in one place and do not spread throughout the body. They may be surrounded by a distinct boundary or capsule (**encapsulated**).
* In **malignant** tumours groups of cancer cells may break off from the main (primary) tumour and spread around the body, where they can grow into other (secondary) tumours. Malignant tumours are less likely to have a distinct boundary or capsule around them. Malignant tumours are usually much more dangerous.

Question

5 The following diagram shows a cancer tumour.

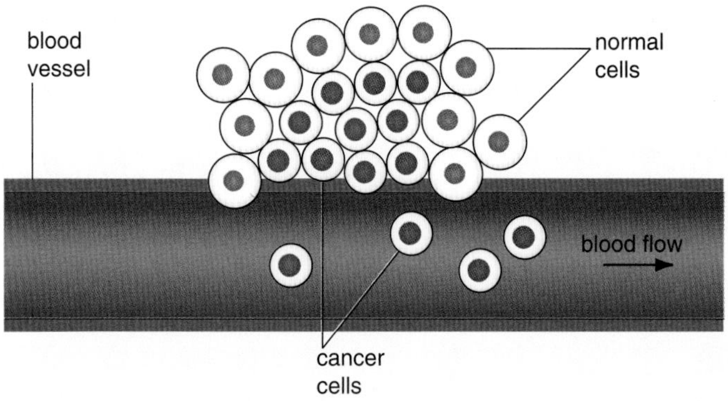

Figure 10.4

a) State one difference between the cancer cells and the normal cells in the diagram.

b) Give two pieces of evidence from the diagram that suggests that this is a malignant tumour.

c) Explain why it is important to find and treat malignant tumours early in their development.

The importance of detecting cancer early

It is very important to detect cancer early. If the cancer is detected early the tumour will be smaller and therefore likely to have caused less damage to the body. It is also very important to detect a malignant tumour before it spreads to other parts of the body.

Early detection often involves **screening programmes**. There are screening programmes for women for both breast and cervical cancer. In these screening programmes women of a certain age and/or with a particular medical history are invited periodically to a medical centre in order to identify if any cancerous cells are present. Screening for skin cancer also helps identify its presence before it spreads too far. Men are advised to check regularly for the early signs of testicular cancer.

Treating cancer

Once cancer is diagnosed there are a number of treatment options. Sometimes more than one option is required. The option(s) used depend on many factors including the type and location of the cancer and the extent of spread through the body. The main options for treatment are:

Surgery
This involves removing the cancer cells from the body. Surgery will be less effective if the cancer has spread throughout the body or the tumour is in an inaccessible part of the body.

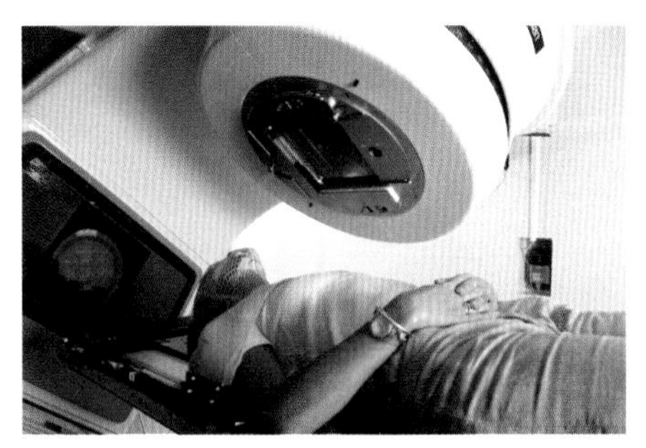

Figure 10.5 A patient receiving radiotherapy

Radiotherapy
X-rays or other forms of radiation can be used to kill the cancer cells. Sometimes other (normal) cells can be affected as well.

Chemotherapy
Chemotherapy involves using a drug or a range of drugs to kill the cancer cells. There are a range of side-effects with chemotherapy, including other (normal) cells being affected, and hair loss.

Finding out more about cancer in Northern Ireland

Search
▶ cancer + Northern Ireland

You can find out more about how the incidence of lung, cervical and skin cancer in Northern Ireland is changing. A very good source is the internet. If you search the internet using multiple-search criteria such as 'Lung cancer' + 'Northern Ireland' or 'Skin cancer' + 'Northern Ireland' or 'Cervical cancer' + 'Northern Ireland', you will find a range of sites that give you good information. Other sources include cancer charities, leaflets from GP surgeries and the occasional article in local newspapers.

Meiosis

Meiosis is another type of cell division. It only takes place in the sex organs (e.g. testes and ovaries) during the production of gametes (i.e. sperm or eggs). The purpose of meiosis is to produce gametes with half the number of chromosomes of all the other (non-gamete) cells in the body. As meiosis halves the chromosome numbers in the daughter cells it is also known as **reduction division.**

Most human cells have 46 chromosomes, arranged in 23 pairs, but the sperm and eggs that are produced by meiosis have only 23 chromosomes. It is not just any 23 chromosomes from the 46 but one chromosome from each pair that passes into each gamete. It could be either chromosome of a particular pair that passes into a particular gamete, and there are 23 pairs of chromosomes in total, so there are millions of potential chromosome combinations in any gamete, i.e. 2^{23} possibilities.

This random **independent assortment** of chromosomes in meiosis at gamete formation gives unique gametes (the chance of any two gametes being identical is so small as to be virtually impossible) and so helps to produce variation in offspring. Figure 10.6 summarises the role of meiosis and the random nature of fertilisation itself in producing variation in living organisms.

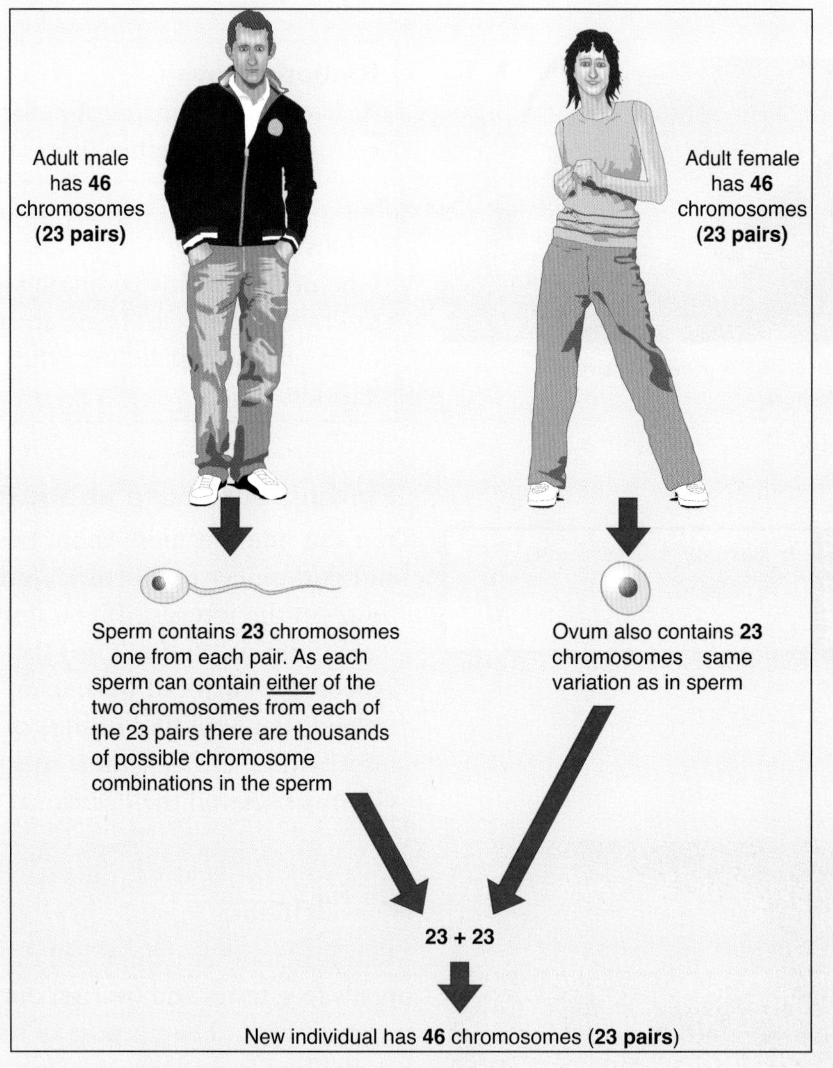

Figure 10.6 Chromosomes in human reproduction

The chromosome number in the gametes is referred to as the **haploid** number (i.e. 23 in humans). The normal number in an organism is called the **diploid** number (i.e. 46 in humans). Clearly the roles of **fertilisation** include restoring the diploid number in the offspring and combining the different arrangements of chromosomes produced during the process of meiosis.

The differences between mitosis and meiosis

Figure 10.7 summarises the differences between mitosis and meiosis.

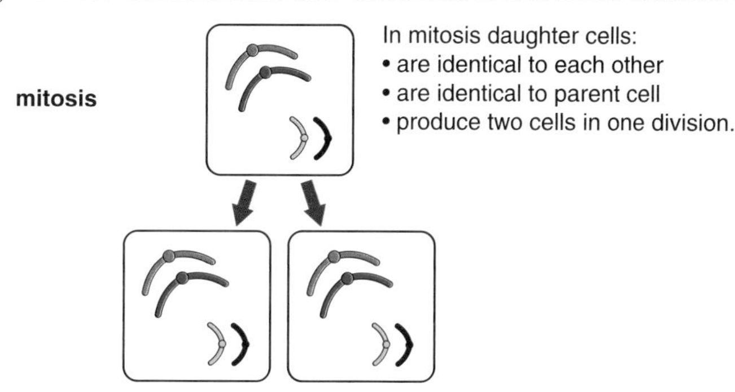

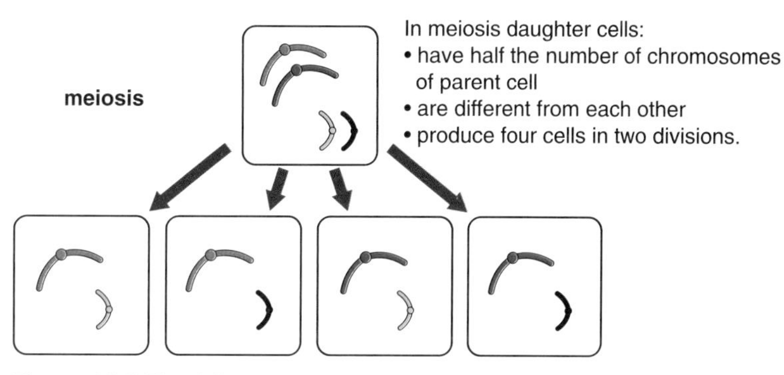

Figure 10.7 The differences between mitosis and meiosis (only two pairs of chromosomes are shown)

We are all aware that young animals and plants resemble their parents. Sometimes the family likeness is so great that it is easy to match offspring with their parents. The passing on of family characteristics from parents to offspring is called **inheritance**. **Genetics** is the scientific study of inheritance.

Gregor Mendel – the founder of genetics

Figure 10.8 Gregor Mendel

Much of our understanding of genetics is based on the work carried out by Gregor Mendel. Mendel was born in Moravia (now the Czech Republic) in 1822 and as a young man joined the church. As a monk in a large monastery he developed an interest in the breeding of the garden pea, plants that were common in the monastery garden. Mendel noticed that the garden pea had many characteristics that varied from plant to plant. These characteristics included pea shape and pea colour. The peas in the garden were either green or yellow and they could be round or wrinkled. Mendel carried out a range of breeding experiments in which he **crossed** (mated) plants carrying particular contrasting characteristics that he was interested in. By careful observation of the offspring produced, he was able to draw conclusions about the nature of inheritance.

One characteristic of pea plants that Mendel was interested in was plant height. Pea plants occur in their normal tall form or in a much shorter dwarf variety. One breeding cross that Mendel carried out was a cross between tall and dwarf plants. Before he carried out this cross he allowed the tall plants to breed with each other for a period of time to ensure they always produced tall plants. He did the same with the dwarf plants by allowing only dwarf plants to breed together until he was sure they could only produce dwarf offspring. The parent plants he used were then referred to as **pure breeding**.

When Mendel crossed the tall plants with the dwarf plants (the parental generation) he found that all the plants in the first, or $\mathbf{F_1}$ **generation** (the offspring) were tall. However, when he crossed these F_1 plants with other F_1 plants their offspring (the second or $\mathbf{F_2}$ **generation**) were a mixture of tall and dwarf plants. Furthermore, as he carried out many crosses that produced hundreds of F_2 plants he worked out that approximately 75% of these were tall with 25% dwarf.

Explanation of the monohybrid cross

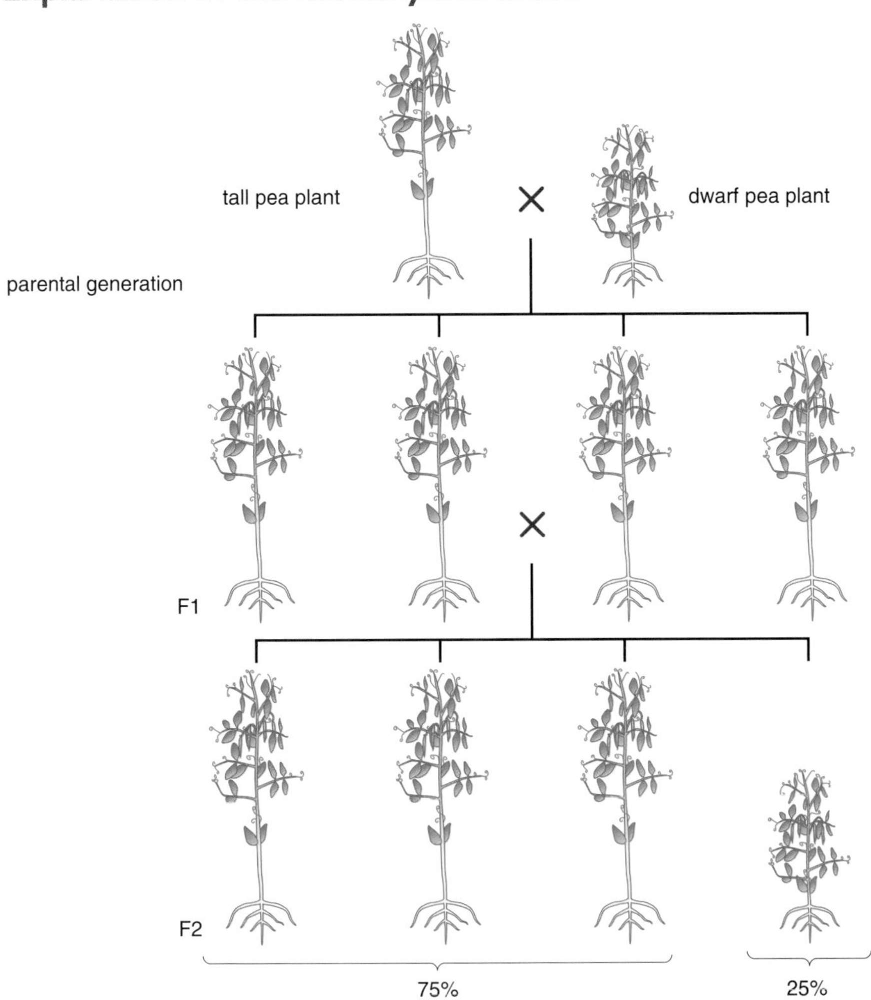

Figure 10.9 Mendel's results

As only one factor was considered in this cross (height of pea plants) it is referred to as a **monohybrid** cross. Mendel decided to give the characteristics he was observing symbols. He used the symbol **T** for the tall plants and the symbol **t** for the dwarf plants. He used a capital letter for the tall state, as it appeared to dominate the dwarf condition. Mendel suggested that there was some factor for tallness in the tall plants and an alternative factor for dwarfness in the short plants. We now know that Mendel's 'factors' are **genes** and that they are carried on the chromosomes. As chromosomes occur in pairs the genes also occur in pairs as shown in Figure 10.10. The two contrasting forms of a gene, i.e. T and t, are called **alleles**. Alleles are different forms of the same gene.

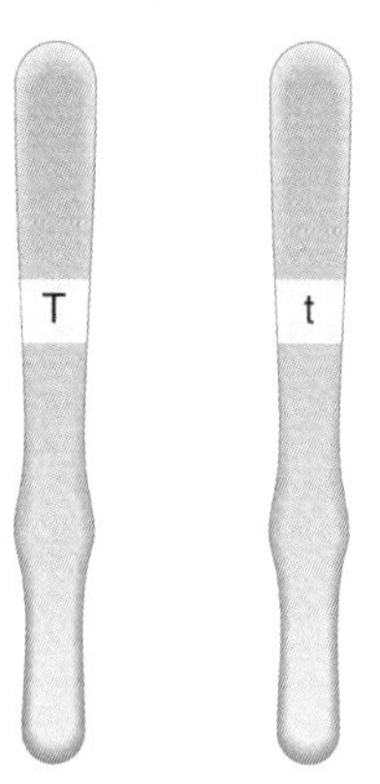

In this example the two alleles of the gene are different. The individual is heterozygous for the characteristic concerned.

Figure 10.10 Chromosomes showing the position of one gene on a pair of chromosomes

Alleles occur in the same position on the chromosome. As the parental plants were pure breeding, Mendel suggested the tall plants only carried the tall factors (genes) and the dwarf plants only carried the dwarf factors. These plants containing only one type of allele are **homozygous** (TT or tt). When both types of allele are present the individual is **heterozygous** (Tt).

The paired symbols used in genetics are referred to as the **genotype** and the outward appearance (tall or short) is the **phenotype**.

Mendel also deduced that when gametes are produced only one factor from each parent passes on to the offspring. This is fully explained by our understanding of meiosis, as we know that only one chromosome, and therefore one allele, of each pair can pass into a gamete.

This was Mendel's **law of segregation**. The two members (alleles) of each pair of genes separate during meiosis, with only one of each pair being present in a gamete.

The F_1 plants in our cross must have received one T allele from their tall parent and one t allele from the dwarf parent. The F_1 plants were therefore Tt (heterozygous). Although all these plants contained both the T and the t allele they were tall. This can be explained by considering the T allele as being **dominant** over the **recessive** t allele. The recessive condition will only be expressed, or visible, in the phenotype when only recessive alleles are present in the genotype (i.e. tt).

Figure 10.11 shows that when the F_1 plants were interbred a ratio of 3:1 (tall:dwarf) was produced.

This ratio is achieved because the two alleles (T and t) of one parent will be produced in equal numbers during meiosis and they have an

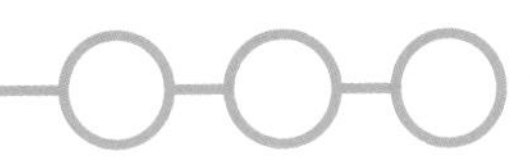

equal chance of combining with the T or the t allele produced by the other parent during fertilisation.

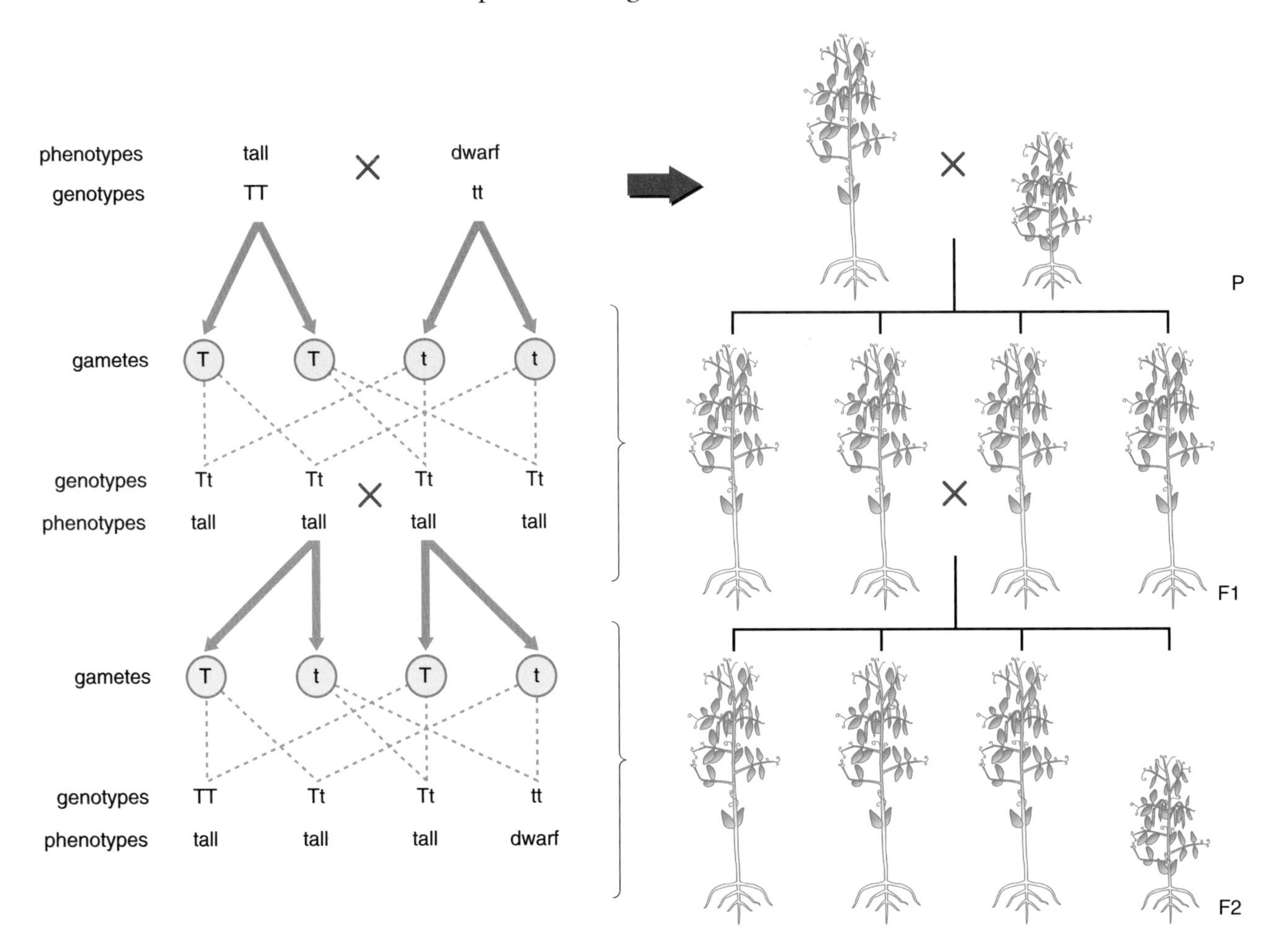

Figure 10.11 Explaining Mendel's results

Mendel completed his work without being aware of the existence of chromosomes or genes. He also had no knowledge of chromosome behaviour in meiosis. Although he published his work in 1866 it did not get the credit it deserved and remained largely ignored. It was only at the start of the twentieth century when chromosomes were discovered under the microscope that the significance of his findings was appreciated. Although additional research and knowledge has increased our understanding of genetics in recent times, it is important to note that this knowledge has built on Mendel's findings as opposed to contradicting them.

Although the inheritance of most human characteristics is complex, often involving many genes and even many chromosomes for a single characteristic, some human features do show monohybrid inheritance. Examples include eye colour and the ability to roll the tongue.

When completing genetic diagrams it is helpful to use a small grid called a **Punnett square**. Figure 10.12 uses Punnett squares to show examples of some monohybrid crosses that can occur in humans.

mother's gametes

father's gametes	B	B
b	Bb	Bb
b	Bb	Bb

100% of offspring have brown eyes
100% of offspring are heterozygous Bb

mother's gametes

father's gametes	B	b
b	Bb	bb
b	Bb	bb

50% of the offspring are heterozygous Bb
and will be expected to have brown eyes

50% are homozygous bb and will
be expected to have blue eyes

mother's gametes

father's gametes	B	b
B	BB	Bb
b	Bb	bb

So there is a 75% chance of these two heterozygous
brown-eyed parents having a brown-eyed child, and
a 25% chance of having a child with blue eyes. This
gives the ratio 3:1 brown:blue.
This percentage chance applies at each conception,
therefore it is possible for all the children to have blue eyes.

Figure 10.12 The inheritance of eye colour in humans (B = brown, b = blue)

Some important points about genetic crosses

* Ratios will only be accurate when large numbers of offspring are produced. This is because it is totally random which gametes, and therefore alleles, fuse during fertilisation.
* It is common practice to use the same letter for both the dominant and recessive alleles, with the dominant allele being the capital and the recessive allele written in lower case.
* If a 3:1 ratio is present in the offspring of a particular cross, both of the parents involved will be heterozygous for the characteristic being considered.
* If a 1:1 offspring ratio is produced in a cross, one parent will be heterozygous and the other homozygous recessive.

The test cross (back cross)

A tall pea plant can be either homozygous (TT) or heterozygous (Tt). Both genotypes give exactly the same phenotype. Sometimes, in agriculture or in the breeding of domestic animals, it is important to know the genotype of a particular animal or plant that is showing the dominant phenotype. To identify the unknown genotype a **test** or **back cross** is carried out.

The animal or plant in question is crossed with a homozygous recessive individual. If the offspring are produced in sufficient numbers it is possible to identify the unknown genotype.

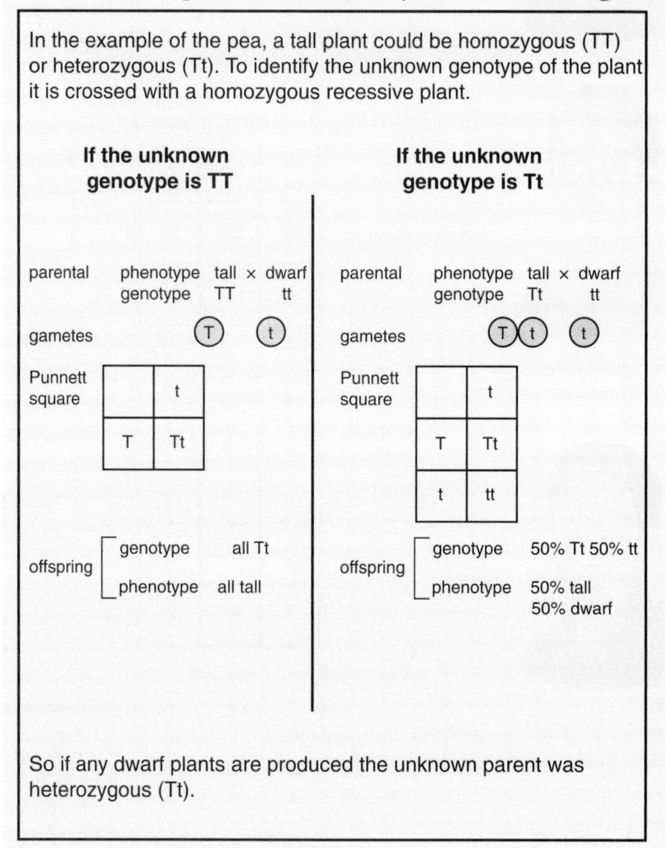

Figure 10.13 The test cross

Pedigree diagrams

A pedigree diagram shows the way in which a genetic condition is inherited in a family or group of biologically related people. Figure 10.14 is an example of a pedigree diagram showing how the condition cystic fibrosis is inherited. Cystic fibrosis is a medical condition caused by a recessive allele.

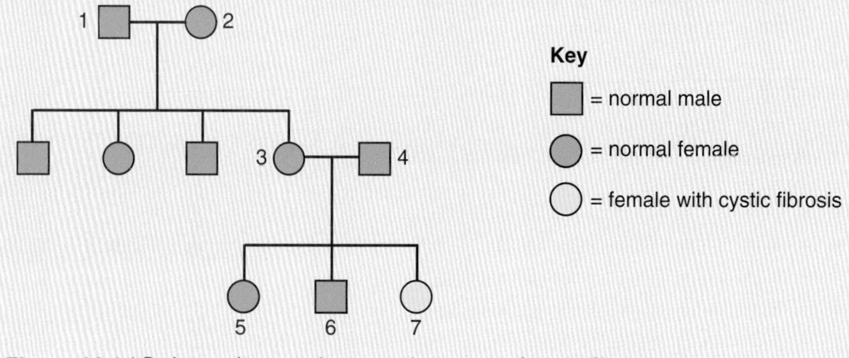

Figure 10.14 Pedigree diagram showing inheritance of cystic fibrosis

In this diagram one of the grandchildren (7) has cystic fibrosis. It is possible to use the information provided to work out the probability of other children having the condition. Genetic counsellors often construct pedigree diagrams and use them to advise parents who have a genetic condition or who may be carriers.

Questions

6 Use Figure 10.14 and your knowledge to answer the following questions. Let C = normal allele; c = cystic fibrosis allele.
  a) What is the genotype of the child (7) with cystic fibrosis? Use the symbol c to represent the cystic fibrosis allele.
  b) What are the genotypes of the parents of child 7 (3 and 4)?
  c) What are the possible genotypes for the brother and sister of the child with cystic fibrosis (5 and 6)? Explain your answer.
  d) What is the probability that the next child of these parents will be a boy with cystic fibrosis?
  e) What can you say about the genotypes of the grandparents of child 7?

Pedigree diagrams can be used in any type of genetic cross but they are obviously very valuable in tracing and predicting harmful genetic conditions.

Sex determination in humans

Sex in humans is another characteristic that is genetically determined. Humans have 46 chromosomes in each cell (except gametes) consisting of 22 pairs of normal chromosomes and one pair of **sex chromosomes**. The sex chromosomes determine the sex of each individual. Males have one X and one Y sex chromosome whereas females have two X chromosomes. A complete set of chromosomes is known as a **karyotype**. Figure 10.15 shows the karyotypes of the human males and female.

During meisosis the female will provide one X chromosome for each egg (ovum), but half the male's sperm will have an X chromosome and half will have a Y chromosome. As there will be an equal chance of an X or a Y chromosome from the male being involved in fertilisation there will be equal numbers of males and females produced. Again, the random nature of fertilisation must be emphasised. We all know of large families consisting of only sons or only daughters. Work out the chance or probability of parents having five children, all of whom are male.

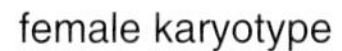

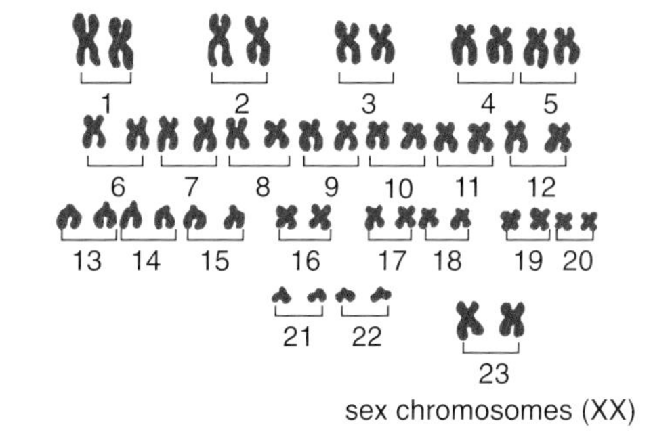

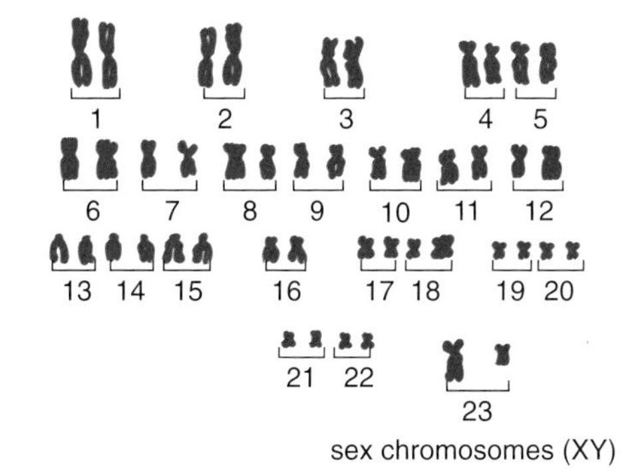

Figure 10.15 The complete set of human chromosomes

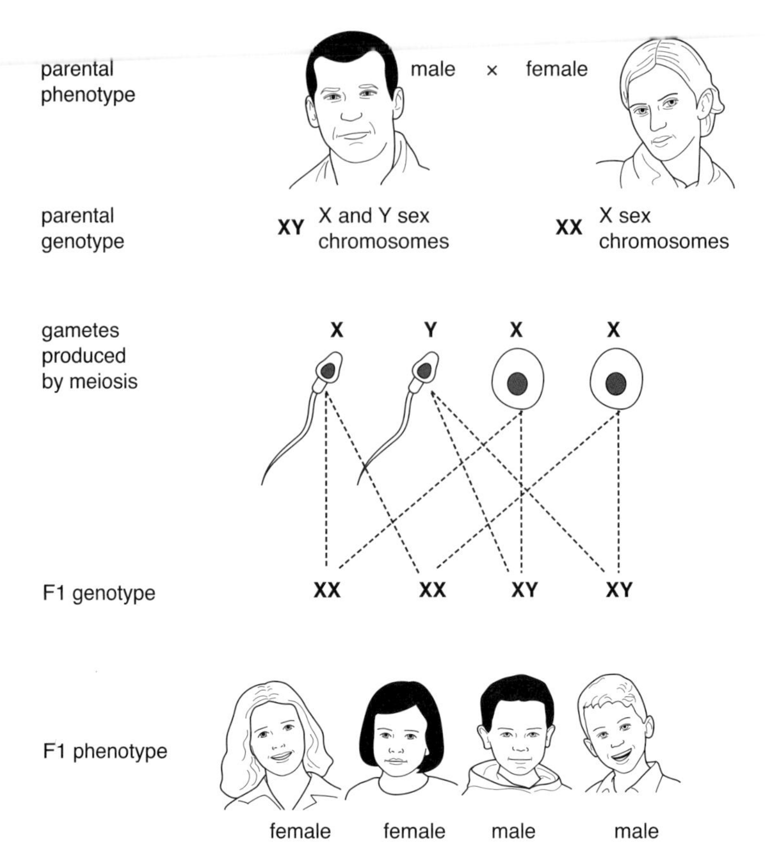

Figure 10.16 Human sex chromosomes

Sex-linkage

The X and Y chromosomes are not only responsible for sex determination. They also have genes that code for a number of body functions. Each of the 22 normal (non-sex) pairs of chromosomes has the same gene present on both chromosomes at the same position. The alleles may be different (alleles for blue or brown eyes) but the gene (gene for eye colour) is present on both. However, in the sex chromosomes the X is much larger than the Y and carries genes that are not present on the Y.

This is particularly important in males as they only have one X chromosome. Therefore any recessive allele carried on an X chromosome in a male will show its effect in the phenotype – there is no dominant allele to mask its effect, as is the situation with females who have two X chromosomes. Haemophilia and red-green colour blindness are sex-linked conditions that are almost exclusively found in males. Females seldom show sex-linked conditions but they are often carriers. In sex-linked conditions, carriers are females who have one dominant and one recessive allele on their X chromosomes. In the female the recessive allele does not affect the phenotype as it is masked by the dominant allele.

Haemophilia is a condition where individuals who are only carrying the recessive allele are unable to make all the products

required to clot their blood. Individuals with red-green colour blindness are unable to distiguish between the colours red and green.

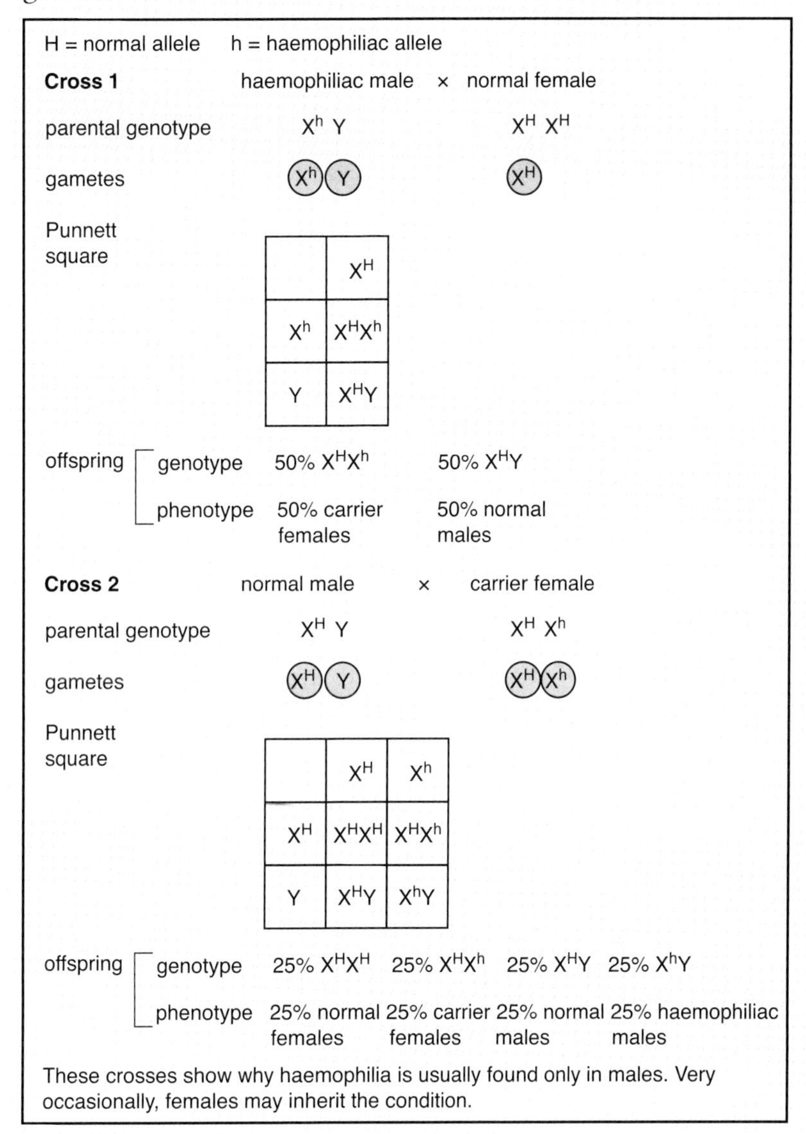

Figure 10.17 The inheritance of haemophilia

Exam questions

1 The diagram shows how plants can be produced by the process of asexual reproduction.

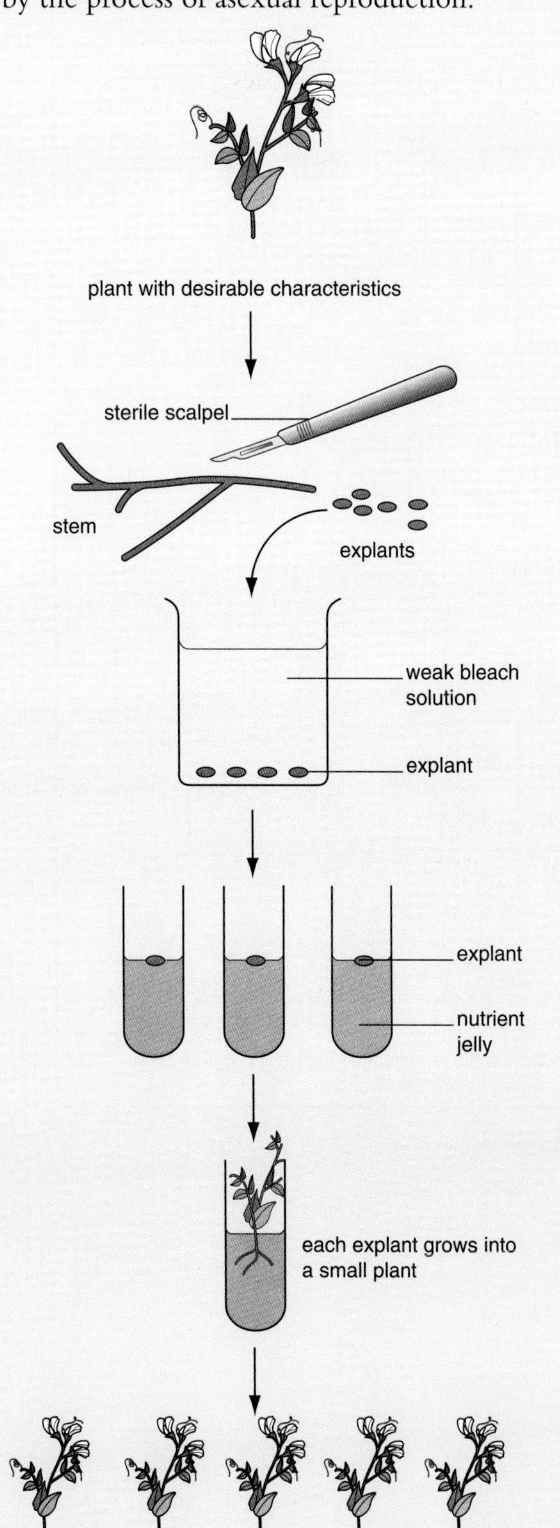

a) Name the process of asexual reproduction illustrated. *(1 mark)*

b) How are the explants obtained? *(1 mark)*

c) Suggest why the explants are placed in a weak bleach solution. *(1 mark)*

d) The explants are placed in agar so that they grow into small plants. Name **one** substance that would need to be present in agar. *(1 mark)*

e) Name the type of cell division involved in producing these plants. *(1 mark)*

2 a) The diagrams show a cross-section through the bladder of a healthy man and one with prostate cancer.

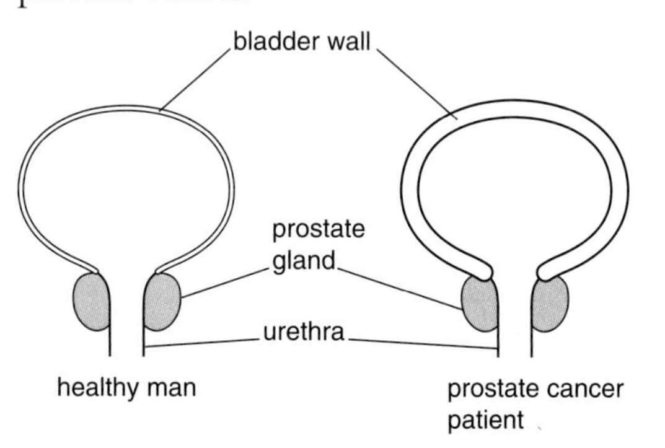

i) Give **one** way in which the cancer patient's bladder differs from that of a healthy man. *(1 mark)*

ii) What is cancer? *(2 marks)*

b) Men can be screened for prostate cancer. Describe how this reduces deaths due to prostate cancer. *(2 marks)*

3 The diagram shows a human life cycle.

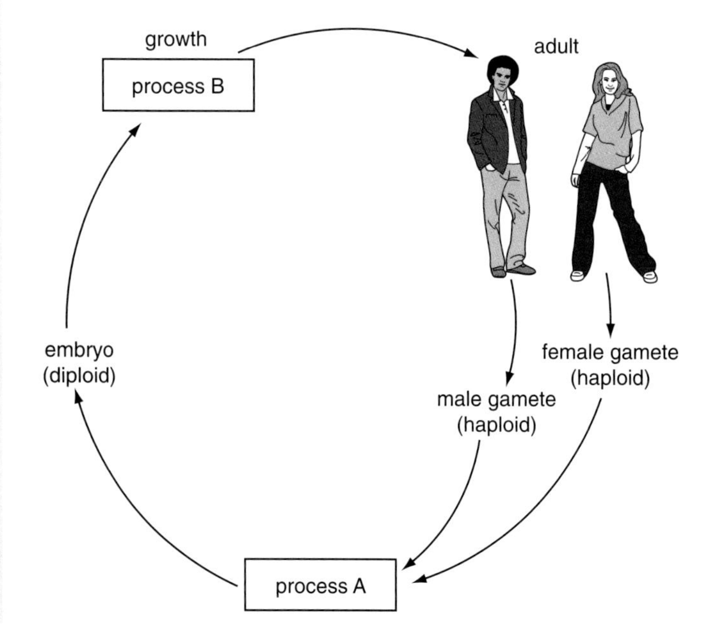

a) Where in the life cycle does meiosis occur? *(1 mark)*

b) Name the processes A and B. *(2 marks)*

c) The diagram shows a cell dividing by process B.

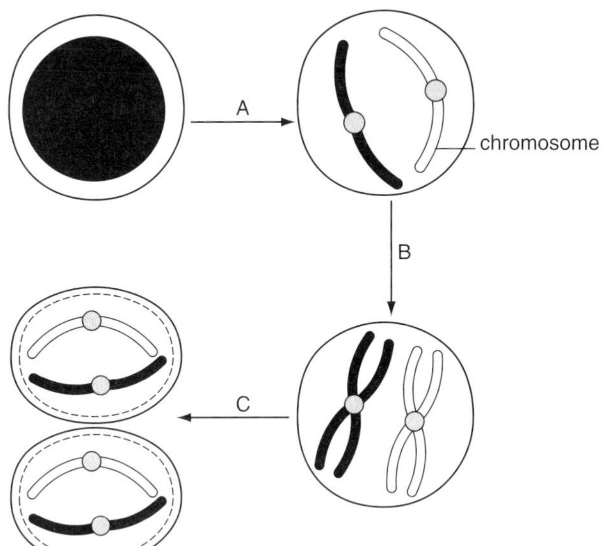

Use the diagram to copy and complete the table below.

Stage	Description of stage	
A		*(2 marks)*
B		*(2 marks)*
C		*(2 marks)*

4 The diagram shows the chromosomes in an animal cell dividing by meiosis.

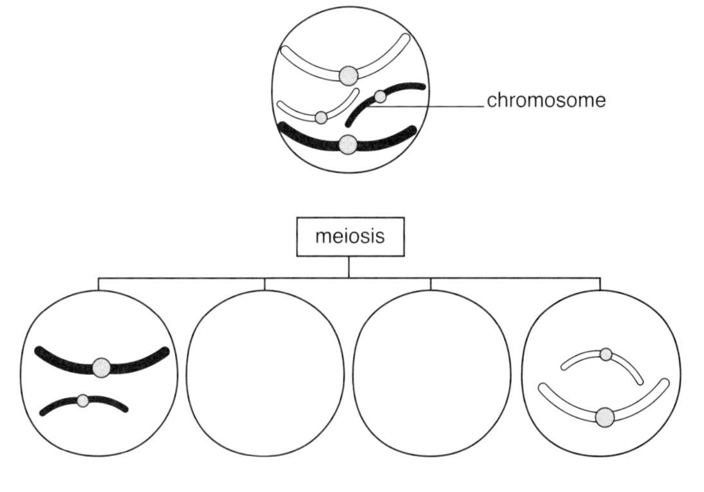

a) Copy and complete the diagram by drawing, in the remaining cells, the other chromosome assortments that would be produced. *(2 marks)*

b) Describe the change in the number of chromosomes during meiosis. *(2 marks)*

c) Where are chromosomes found in a cell? *(1 mark)*

d) Where in the body does meiosis occur? *(1 mark)*

e) Give **two** differences between mitosis and meiosis. *(2 marks)*

5 a) The diagram shows an example of a particular type of cell. This type of cell divides to produce gametes. Copy and complete the diagram to show the result of such cell division. *(3 marks)*

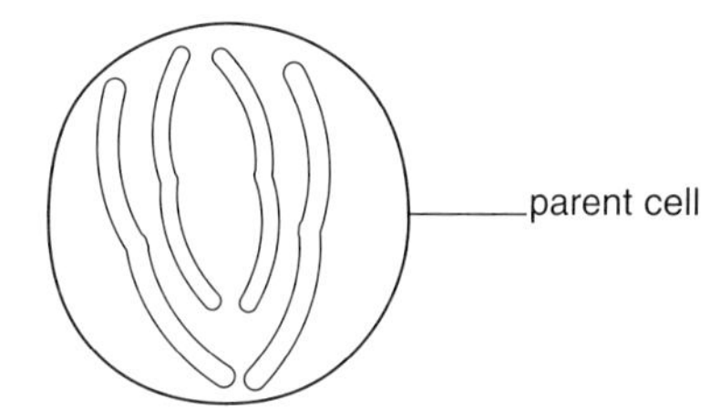

b) Name the type of cell division that produces gametes. *(1 mark)*

c) The parent cell has a diploid number of chromosomes. What term describes the number of chromosomes in the gamete? *(1 mark)*

6 a) Brown eye colour is dominant to blue eyes. Mary has blue eyes and her partner Tony has brown eyes. Their son John has blue eyes. Copy and complete the diagram below to show how John inherits blue eyes.
Use symbols: B = brown allele b = blue allele.

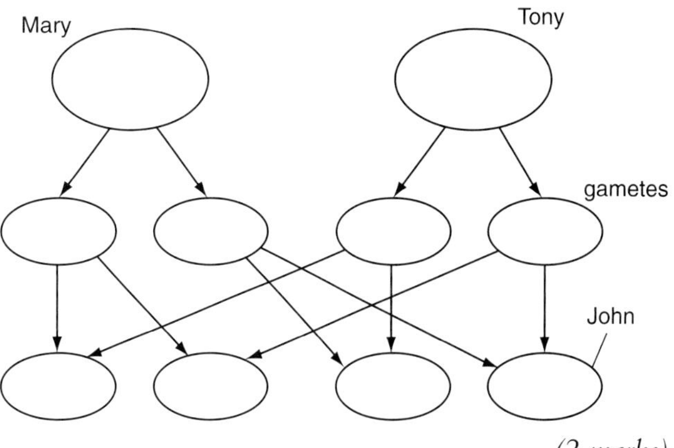

(2 marks)

b) What is the probability (chance) of Mary and Tony's next child having brown eyes? *(1 mark)*

c) Using eye colour as an example, explain the term **heterozygous**. *(2 marks)*

d) Use your knowledge of sex determination in humans to describe why equal numbers of boys and girls are born. *(3 marks)*

7 Pea plants can produce peas that are wrinkled or smooth.
The gene (allele) for wrinkled is dominant to the gene (allele) for smoothness.
Let R represent the gene (allele) for wrinkled peas.
Let r represent the gene (allele) for smooth peas.

a) A plant breeder crossed two heterozygous pea plants.

i) Use a Punnett square to show the possible genotypes of the offspring of this cross. *(3 marks)*

ii) Give the phenotypes of the offspring and the ratio of the phenotypes. *(2 marks)*

b) The diagram shows a cell with chromosomes.

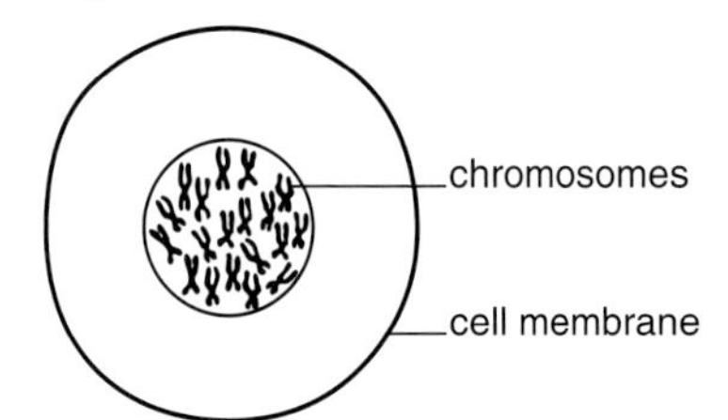

i) Where in the cell are the chromosomes located? *(1 mark)*

ii) What are short sections of chromosomes called? *(1 mark)*

iii) What chemical are chromosomes made from? *(1 mark)*

c) The following diagram shows how the sex of a child depends on the chromosomes it inherits from its parents.

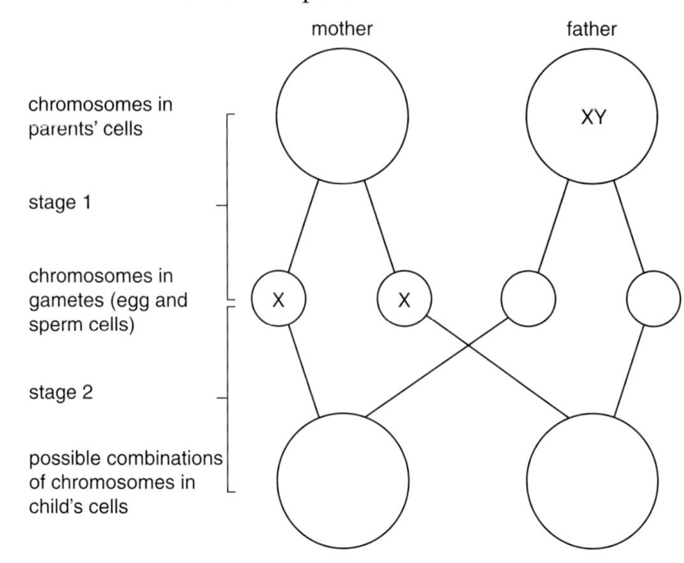

Copy and complete the diagram above to show the sex chromosomes found in each cell. *(3 marks)*

8 a) The diagrams show the results of a breeding experiment with fruit flies.

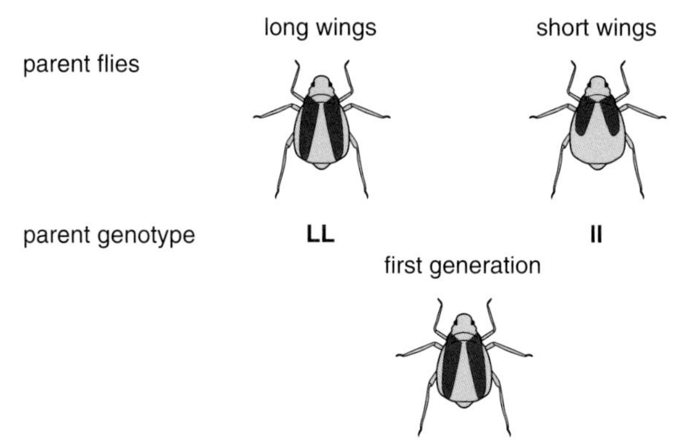

the allele for long wings is **L**
the allele for short wings is **l**

i) Which allele is dominant? *(1 mark)*

ii) Give the genotype of the first generation fly. *(1 mark)*

b) A male and a female from the first generation were crossed.
Copy and complete the Punnett square to show the genotypes of the second generation flies.

		Long wing male gametes	
			l
Long wing female gametes	L		

(2 marks)

c) Geneticists wanted to determine the **genotype** of a long wing fly.
Describe the procedure they would have to follow to determine its genotype.
In this question you will be assessed on using your written communication skills including the use of specialist science terms. *(6 marks)*

9 Haemophilia is a genetic disorder which is sex-linked.
The diagram shows a family tree.

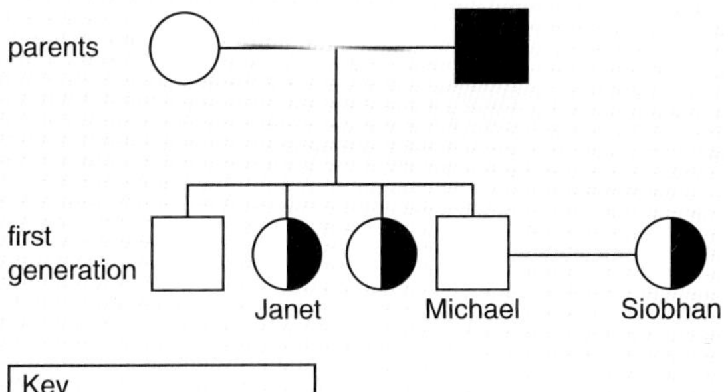

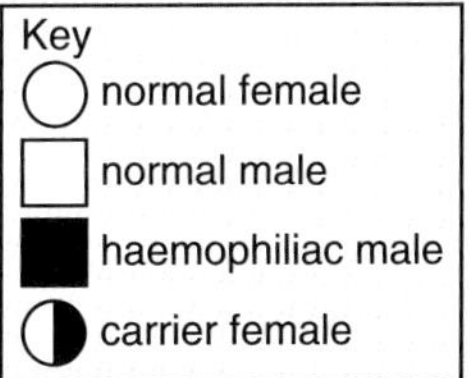

Let X^H represent a normal X chromosome
X^h represent an X chromosome carrying the haemophiliac allele
Y represent a Y chromosome.

a) Give the genotypes of the parents. *(2 marks)*

b) Give the phenotype and genotype of Janet. *(2 marks)*

c) Explain how Michael is a normal male even though his father is a haemophiliac. *(2 marks)*

Michael and Siobhan have children.

d) What proportion of the children will have haemophilia? *(1 mark)*

e) Why are there fewer haemophiliac females than males? *(1 mark)*

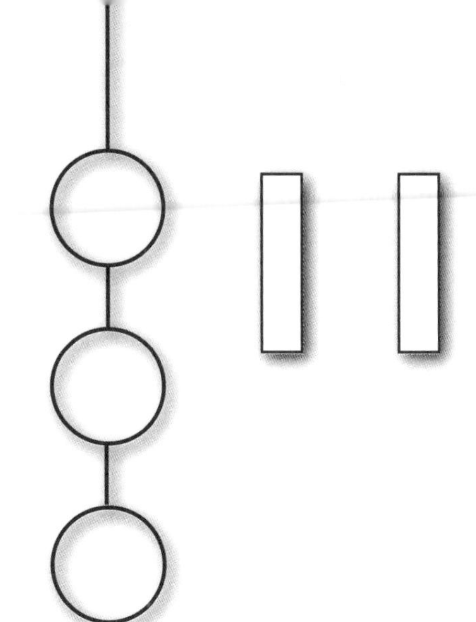

11 Reproduction, Fertility and Contraception

Living organisms need to be able to reproduce or they would no longer exist. Humans, as with most animals, carry out **sexual reproduction**.

▶ The sperm and eggs – very specialised cells

Sperm and egg cells are **gametes** or sex cells. They are **haploid** cells – this means they have only half the number of chromosomes that normal body cells have. When a sperm cell and an egg cell combine during **fertilisation** the normal number of chromosomes (**diploid**) is restored. For example, most human cells have 46 chromosomes, whereas sperm and egg cells each have 23. Sperm and egg cells are made during the process of **meiosis**.

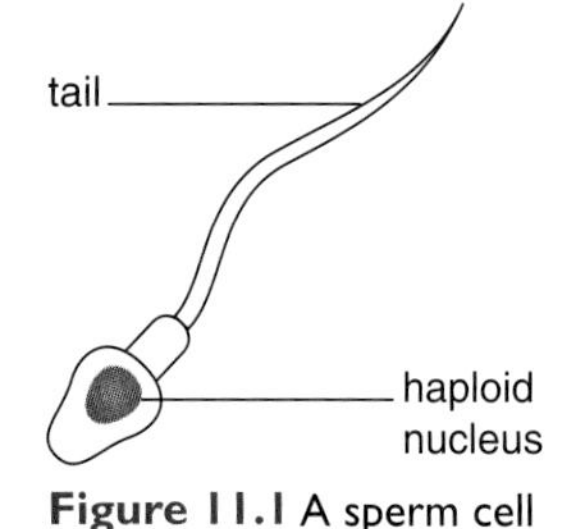

Figure 11.1 A sperm cell

Sperm cells are produced in the testes under the influence of the hormone testosterone. Their adaptations include the haploid nucleus and a tail for swimming.

Egg cells (**ova**) are larger than sperm cells and are produced in the ovaries under the influence of female hormones.

▶ Fertilisation, pregnancy and the development of the baby

If a sperm and an egg (ovum) meet and **fuse** (join) in an **oviduct**, fertilisation will result. Fertilisation involves the haploid nuclei of the sperm and egg fusing and restoring the diploid (normal chromosome number) condition. The fertilised egg becomes the first cell of the new individual, the **zygote**. This cell then divides by **mitosis** and grows into a ball of cells as it travels down the oviduct. The ball of cells becomes an **embryo**, which becomes attached (**implanted**) to the wall of the **uterus**. To enable this to happen the uterus develops a thick lining that holds and nourishes the embryo.

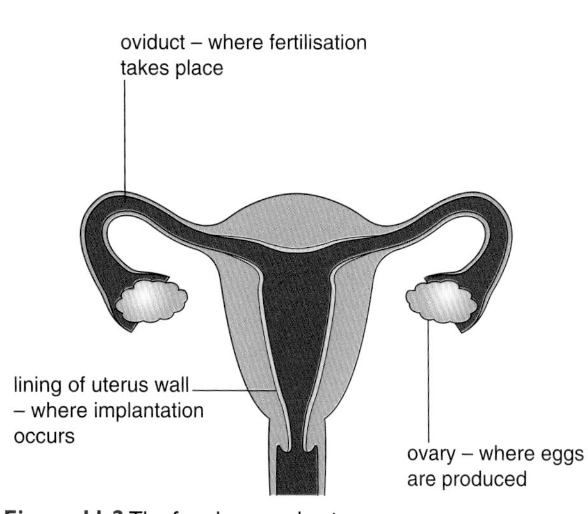

Figure 11.2 The female reproductive system

At the point where the embryo begins to develop in the uterus lining, the **placenta** and **umbilical cord** form. A protective membrane, the **amnion**, develops around the embryo. It contains a fluid, the **amniotic fluid**, within which the growing embryo develops. This fluid cushions the delicate developing embryo, which increasingly differentiates into tissues and organs. The embryo is referred to as a **foetus** after a few weeks when it begins to become more recognisable as a baby.

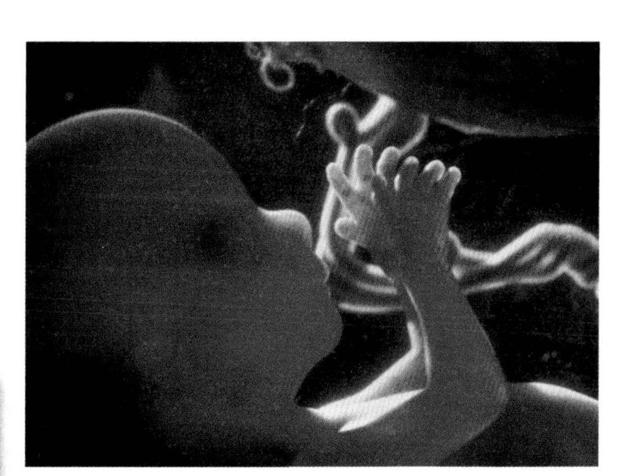

Figure 11.3 Human foetus at four months, showing the umbilical cord and the placenta

Questions

1. Apart from oxygen and glucose, name one other essential substance that will pass from the mother to the foetus during pregnancy.
2. Name two excretory materials that will pass from the foetus to its mother.

Obviously the baby cannot breathe when in the amniotic fluid (its lungs will not be developed enough anyway), so during pregnancy useful materials including oxygen and glucose pass from the mother to the foetus through the placenta and umbilical cord. Waste excretory materials pass from the foetus back to the mother.

The placenta and umbilical cord in more detail

Figure 11.4 shows the very close relationship between the blood vessels of the uterus (which is maternal tissue) and the blood vessels in the placenta.

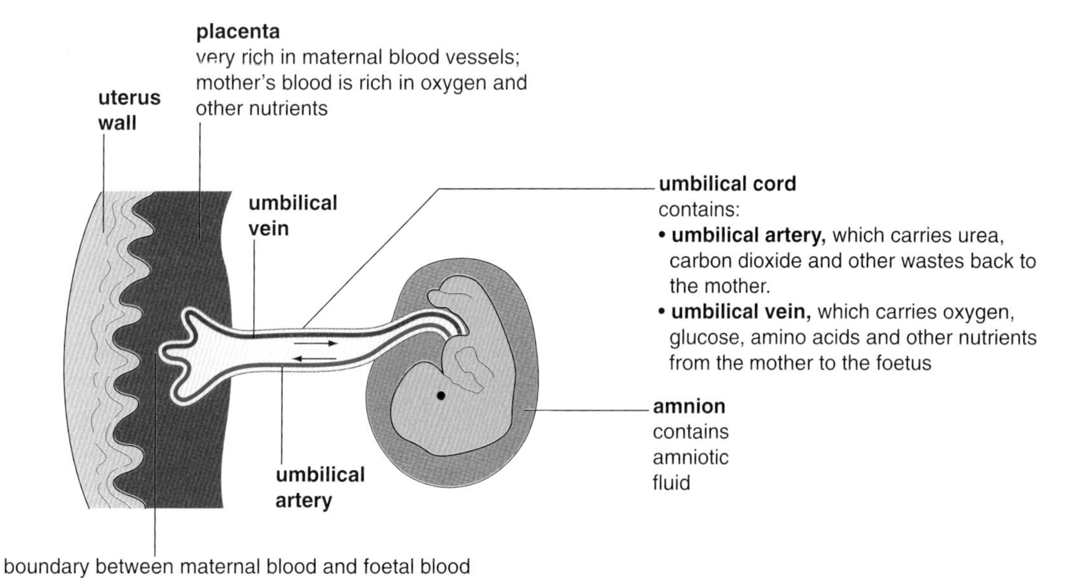

Figure 11.4 The functions of the placenta

The surface area between the uterine wall and the placenta is further increased by small 'villi' in the placenta that extend into the uterine wall.

▶ Contraception – preventing pregnancy

Many people want to have sex but do not want to have children at that particular time. Pregnancy can be prevented by **contraception**.

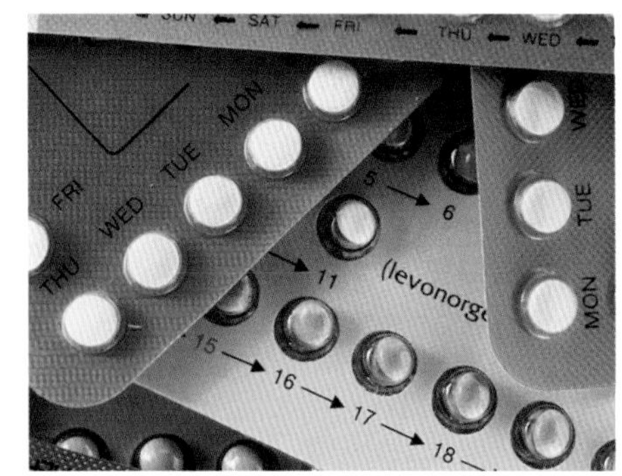

Figure 11.5 Methods of contraception

Methods of contraception

There are three main types of contraception: **mechanical**, **chemical** and **surgical**. Examples of each and an explanation of how they work, together with their main advantages and disadvantages, are given in the table below.

Type	Example	Method	Advantage	Disadvantage
Mechanical (physical)	Condom	Acts as a barrier to prevent the sperm entering the woman	Easily obtained and also protects against sexually transmitted diseases such as AIDS, chlamydia and gonorrhoea	Unreliable if not used properly
Chemical	Contraceptive pill	Taken regularly by the woman and prevents the ovaries from releasing eggs by changing hormone levels	Very reliable	Can cause some side-effects such as weight gain and may increase the risk of blood clots
Surgical	Vasectomy	Cutting of sperm tubes, preventing sperm from entering the penis	Virtually 100% reliable	Very difficult or impossible to reverse
	Female sterilisation	Cutting of oviducts, preventing ova from moving through the oviduct and being fertilised	Virtually 100% reliable	Very difficult or impossible to reverse

Some people are opposed to contraception but may want to *reduce* their chances of having children – often because they have a large family already. They can do this by avoiding having sex around the time when the woman releases an ovum each month – this has been called the **rhythm** or **natural method** of contraception.

Some people choose this method for religious or ethical reasons but it is much less effective as a contraceptive method. In many women the menstrual cycle is irregular, making it difficult to know exactly when an egg is being released.

Questions

3 Why do you think a 25-year-old man who is married with no children is likely to be asked to reconsider his request for a vasectomy?

4 a) Draw a diagram of the female reproductive system. On it label the ovaries, oviducts, uterus, cervix and vagina.

b) On the diagram show where:

i) ova are released

ii) fertilisation will take place

iii) surgical contraception can take place.

Testosterone and oestrogen, and the development of secondary sexual characteristics

Testosterone (produced by the testes in males) and **oestrogen** (produced by the ovaries in females) are important hormones in overall sexual development. One effect they have is the development of the secondary sexual characteristics that are a feature of puberty. The changes that occur in males and females are different but in both sexes they serve to prepare the body for reproduction, both physically and by increasing sexual awareness and drive.

Some of the secondary sexual characteristics produced by testosterone and oestrogen are summarised in the table below.

Males	Females
Body hair and pubic hair develops	Hair grows in pubic regions and in the armpits
The sexual organs (genitals) enlarge	The sexual organs enlarge and the breasts develop
The body becomes more muscular	The pelvis and hips widen
The voice deepens	Menstruation begins
Sexual awareness and drive increase	Sexual awareness and drive increase

▶ The menstrual cycle

The **menstrual cycle** occurs in females from puberty until the end of reproductive life (usually some time between the ages of 45 to 55). The purpose of the menstrual cycle is to prepare the reproductive system for pregnancy by controlling the monthly release of an egg and renewing and replacing the uterine lining. The cycle is controlled by the female hormones (including oestrogen, which is the sex hormone responsible for the secondary sexual characteristics covered above).

In humans the menstrual cycle is approximately 28 days long. The cycle begins with **menstruation** (days 1–5 approx.). This is when the blood-rich uterine lining breaks down and is passed out of the body. After menstruation the uterine lining repairs itself and builds up again in preparation for fertilisation and implantation should it occur (days 6–13 approx.).

At around 14 days into the cycle an egg is released (**ovulation**) by an ovary. By this time the uterine lining has been fully repaired and is ready for pregnancy should it happen. It is only at this time in the cycle that **pregnancy** could occur, as there is an egg present should sperm also be present.

If implantation and pregnancy do occur, an embryo will start to develop. The uterine lining then becomes even thicker and a placenta forms. If pregnancy does not occur, the uterine lining does not develop further after ovulation.

▶ Fertility problems and their treatment

Some people have problems that prevent them having children (fertility problems). Reasons include:

* failure of ovary to produce eggs
* the oviducts may be blocked or twisted, possibly due to infection

* complications of some sexually transmitted infections
* the lining of the uterus does not develop properly to enable implantation to occur
* the vagina may be hostile to sperm entering, e.g. the lining may be too thick or too acidic
* males may not produce enough sperm or the sperm may not be healthy – this can be affected by smoking or taking alcohol in excess
* impotence.

Treatments may include:

Fertility drugs (hormone treatment)
These are given to the woman to increase production of eggs. This may solve the problem if low egg production is the issue but if there are other problems such as blocked oviducts then *in vitro* fertilisation may be necessary.

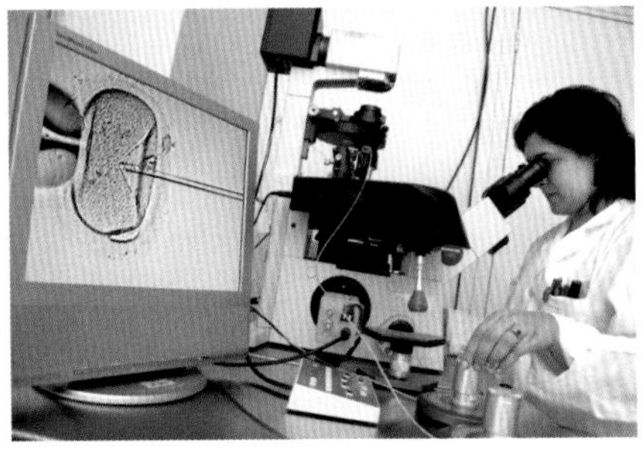

Figure 11.7 *In vitro* fertilisation

***In vitro* fertilisation (test-tube babies)**

* The woman is given fertility drugs so that several eggs are produced.
* These are collected from the ovaries surgically.
* Sperm is donated and the sperm and eggs are mixed in the laboratory.
* Successful embryos are placed in the mother's uterus (she will have undergone hormonal treatment to ensure her uterus lining is ready). If the process is successful an embryo (or possibly more than one) will implant in the uterus lining.
* Usually only a small number of embryos are placed in the mother's uterus to give a balance between ensuring a successful pregnancy and avoiding multiple births.

Search
▶ fertility problems

Fertility research and treatment is a controversial area. It is now possible to screen embryos to check for abnormalities and even to check the sex of embryos before they are placed in a woman's uterus.

Exam questions

1 The diagram shows the chromosomes in a cell. Cells divide by mitosis or meiosis. A, B, C, D and E show daughter cells which *may* be produced as a result of these types of cell division.

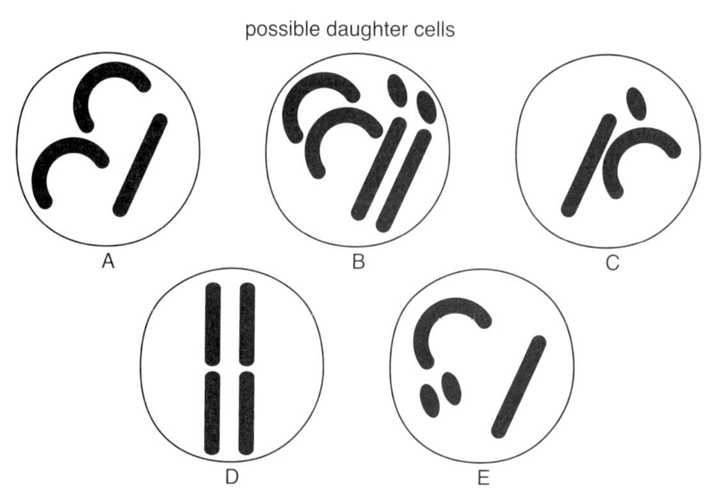

a) Copy and complete the following table by identifying the correct daughter cell for each type of cell division.

Type of cell division	Cell
Mitosis	
Meiosis	

(2 marks)

b) i) Where does meiosis take place in the male reproductive system? *(1 mark)*

ii) Name the hormone that influences this process in the male reproductive system. *(1 mark)*

iii) Fertilisation takes place in the oviduct to form a diploid zygote. What type of cell division must then take place before implantation occurs? *(1 mark)*

c) The diagram shows the female reproductive system.

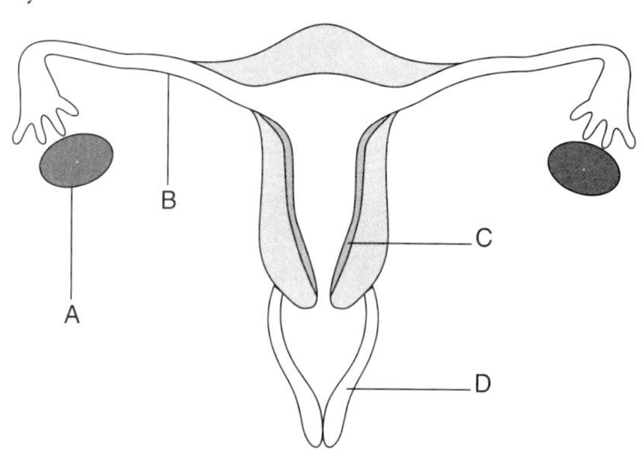

i) In which part, A, B, C or D, does implantation occur? *(1 mark)*

ii) After implantation, what process occurs to produce a variety of tissues and organs, e.g. amnion and placenta? *(1 mark)*

iii) Give the functions of the amnion and placenta. *(2 marks)*

iv) Describe how the placenta is adapted to carry out its function. *(1 mark)*

2 The diagram shows an early human foetus attached to its placenta.

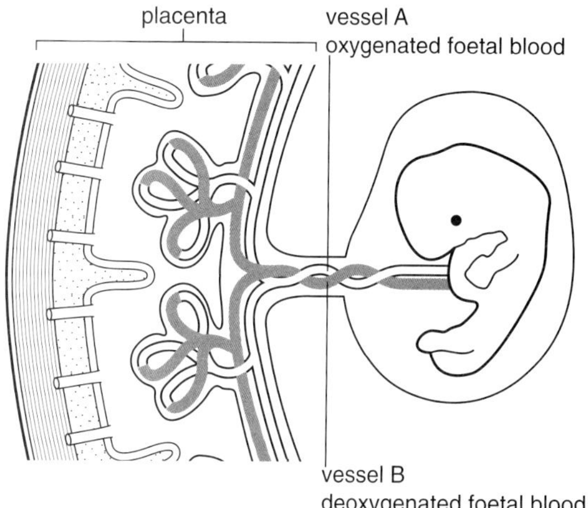

a) In whch direction does the blood flow in vessel A?

b) Name **one** substance, other than oxygen, which passes from the mother's blood to the foetus. *(1 mark)*

c) Give **one** feature of the placenta and explain how it maintains a high rate of exchange of substances between the mother and the foetus. *(3 marks)*

3 The diagram shows the menstrual cycle.

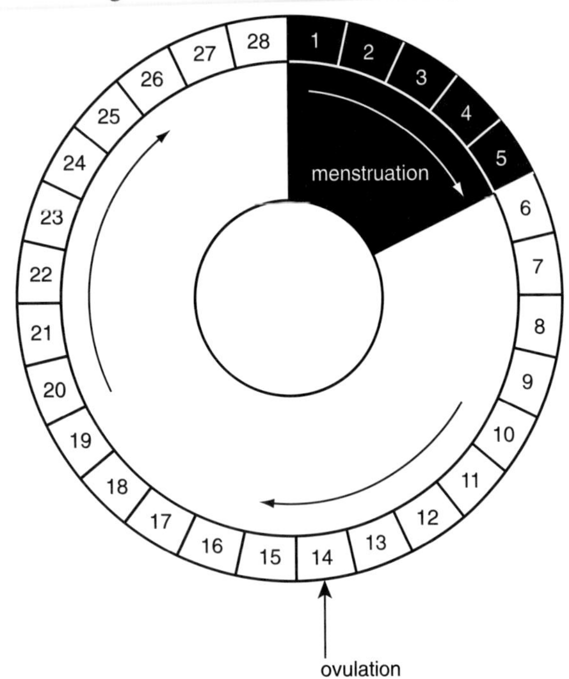

Use the information in the diagram to help describe the changes that occur in the female reproductive system from day 1 of the menstrual cycle leading up to pregnancy.

In this question you will be assessed on using your written communication skills including the use of specialist science terms *(6 marks)*

4 *In vitro* fertilisation is one development that helps infertile couples.

a) Explain how a woman is made to produce a large number of ova at the start of *in vitro* fertilisation. *(1 mark)*

b) Describe how the ova produced are fertilised during *in vitro* fertilisation. *(2 marks)*

c) Describe what must happen to the resulting embryos, if the woman is to become pregnant. *(2 marks)*

d) There is some controversy associated with the use of *in vitro* fertilisation.

i) Give **one** argument for *in vitro* fertilisation. *(1 mark)*

ii) Give **one** argument against *in vitro* fertilisation. *(1 mark)*

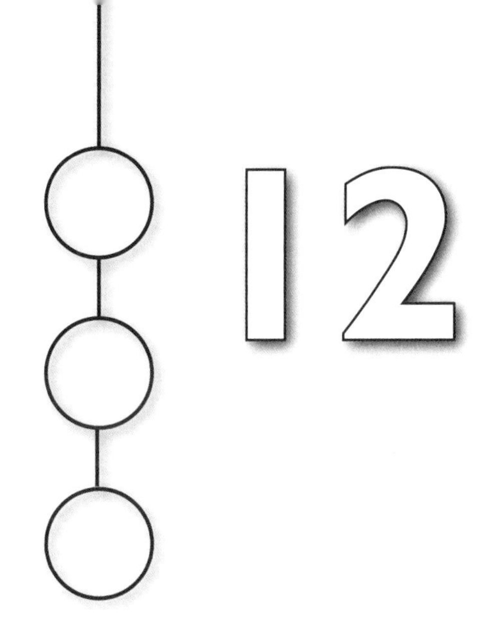

12 Applied Genetics

Mutations

Random changes to the structure or number of chromosomes or genes are called **mutations**. Down syndrome is caused by a change to the *number* of chromosomes. Many other mutations result in a change in gene or chromosome *structure* as opposed to number.

The condition haemophilia is caused by a mutation producing a recessive allele instead of the normal dominant allele.

Most mutations are harmful and they usually arise randomly. However, there are a number of environmental factors that can greatly increase the chances of a mutation occurring. A good example is the link between UV (sun) light and skin cancer. Skin cancer is caused by changes to gene structure.

Questions

1 Suggest why the incidence (occurrence) of skin cancer in people who live in the British Isles has increased dramatically in recent times.
2 Outline the precautionary measures that can be taken to decrease the likelihood of getting skin cancer.

Down syndrome in more detail

Down syndrome is also a human condition. This condition is not caused by a recessive allele but by an error in the formation of gametes. In humans, gametes (sperm and eggs) normally have 23 chromosomes (one from each of the 23 pairs of chromosomes). Occasionally gametes are formed with 24 chromosomes. If the affected gamete is fertilised, resulting in pregnancy, the new individual will have 47 chromosomes in all of his or her cells instead of the normal 46.

There are a large number of conditions that are either caused directly by the chromosomes and genes that an individual possesses (as in the examples above), or where there is at least a strong genetic component. Several of the medical conditions that are discussed in this book are thought to have possible genetic factors. These include diabetes and heart disease. As our knowledge of inheritance and genetics increases it is hoped that we will eventually be able to reduce the occurrence of genetic disease in the population. One way of reducing the occurrence or effects of genetic conditions is genetic screening, which will be discussed later in this chapter.

Figure 12.1 shows a karyotype (a chromosome spread) of an individual with Down syndrome and an individual without the condition.

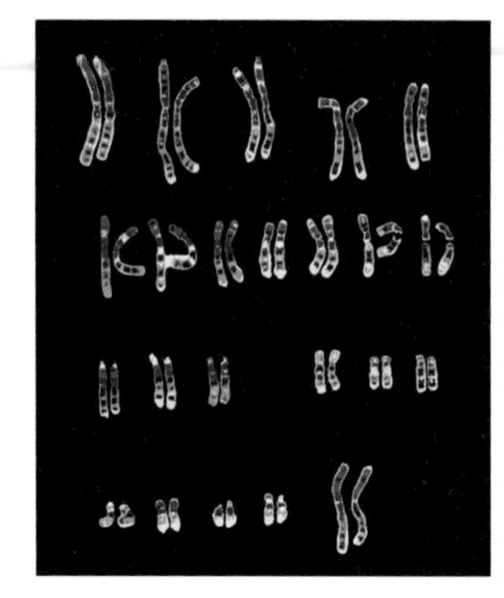

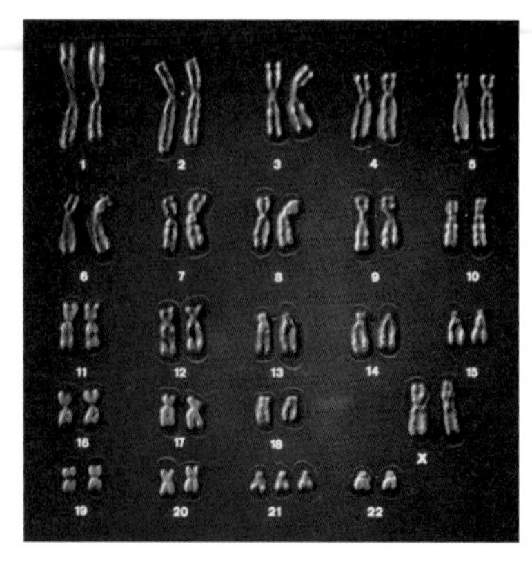

Figure 12.1 Karyotypes of a normal individual (left) and an individual with Down syndrome (right). Note that with Down syndrome there is an extra chromosome 21

Genetic screening

Genetic screening may be used to reduce the incidence of diseases or conditions caused by problems with our chromosomes or genes. It involves testing people for the presence of a particular allele or genetic condition. Whole populations can be tested, or targeted groups where the probability of having (or passing on) a particular condition is high. Genetic screening can be a particular issue for pregnant mothers and their partners.

Genetic screening has been available for a long time for Down syndrome. In screening for Down syndrome, cells can be taken from the amniotic fluid surrounding the baby in the womb and they are then allowed to multiply in laboratory conditions. The chromosomes in the cells can then be examined to see if the developing foetus has the condition (an **amniocentesis** test). This genetic screening is offered to pregnant women in Britain but it is probably more important for older mothers. Genetic screening is also available for cystic fibrosis. Clearly mothers who know that they and/or their partners are carriers for cystic fibrosis also have to consider the potential implications before becoming pregnant.

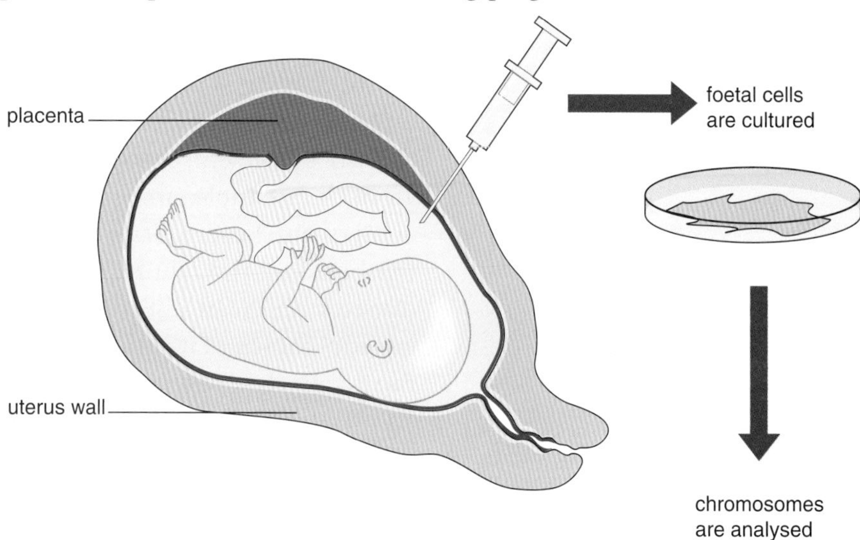

Figure 12.2 An amniocentesis test

Genetic screening – the ethical issues

What are the options if genetic screening shows that a foetus has a condition such as Down syndrome? Should the mother have an abortion? The results of genetic screening and even being a carrier of a genetic condition can create considerable dilemmas for the individuals involved.

This is a good topic for discussion in a class debate. Some of the issues you may wish to consider are highlighted below.

* The ethics of abortion for medical reasons.
* Is there an acceptable risk associated with genetic screening? For example, amniocentesis for Down syndrome screening has a small risk of miscarriage.
* Should parents be allowed free choice whether to screen or not?
* Should you be allowed to screen for the sex of a child? What if it is not the sex you want?

It will soon be possible to screen everyone (whether before birth, in childhood or as an adult) for many different alleles. The information obtained is referred to as a **genetic profile**. Should this information be available to life insurance companies and employers?

* Costs of screening compared to the costs of treating individuals with a genetic condition – should cost be a factor?
* Should genetic screening be extended to more than just serious genetic conditions? What if it can predict life expectancy?

Genetic engineering

The speed of increasing scientific knowledge is such that within 50 years of working out the structure of DNA, scientists were able to manipulate and change the DNA in living organisms. **Genetic engineering** involves taking a piece of DNA, usually a gene, from one organism (the donor) and adding it to the genetic material of another organism (the recipient). Commonly, DNA that codes for the desired product is incorporated into the DNA of bacteria. This is because bacterial DNA is easily manipulated and also because bacteria reproduce so rapidly that large numbers can quickly be produced with the new gene. As a result the bacteria will produce a valuable product, such as a drug or hormone that may be difficult or expensive to produce by other means. Once the new genetic material is built into the bacteria they are allowed to reproduce rapidly in suitable growing conditions (cultured) in special fermenters or bioreactors that maximise the production of the desired product.

One of the best examples of genetic engineering providing essential products for man is in the production of genetically engineered human insulin. Diabetes is becoming increasingly common and as a result many more people require insulin than did in the past. Before the development of genetic engineering the insulin was obtained from the pancreases of domestic animals such as pigs and cattle.

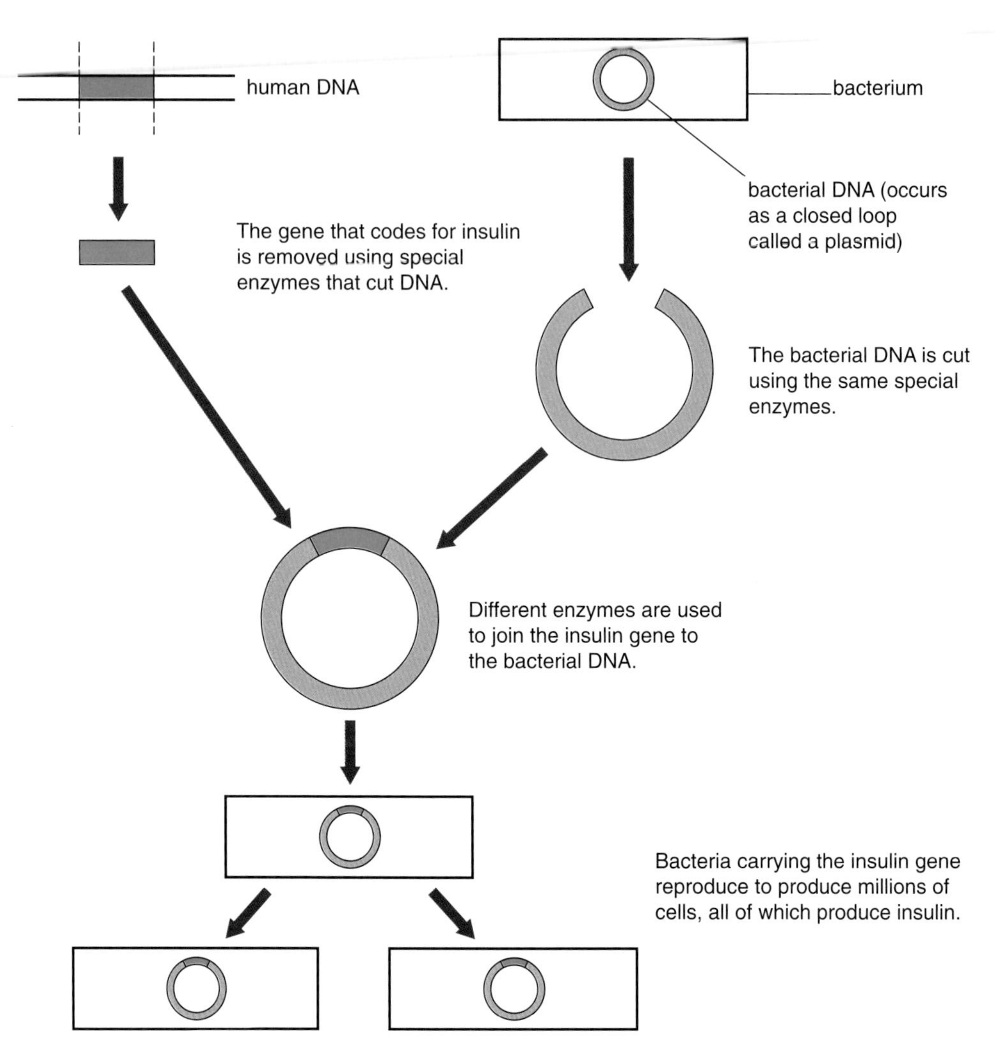

Figure 12.3 Genetic engineering – making human insulin

The enzymes that cut and isolate the human insulin gene and cut the plasmid are called **restriction enzymes**. These cut the DNA in such a way that one of the two strands extends further than the other one. The longer strand will have 'free' exposed bases that are not paired. The key thing is that each restriction enzyme will leave complementary sections of exposed bases in both the plasmid and the human insulin gene so that they can join by base pairing between each other. Not surprisingly the exposed strands of DNA and their bases are called '**sticky ends**'.

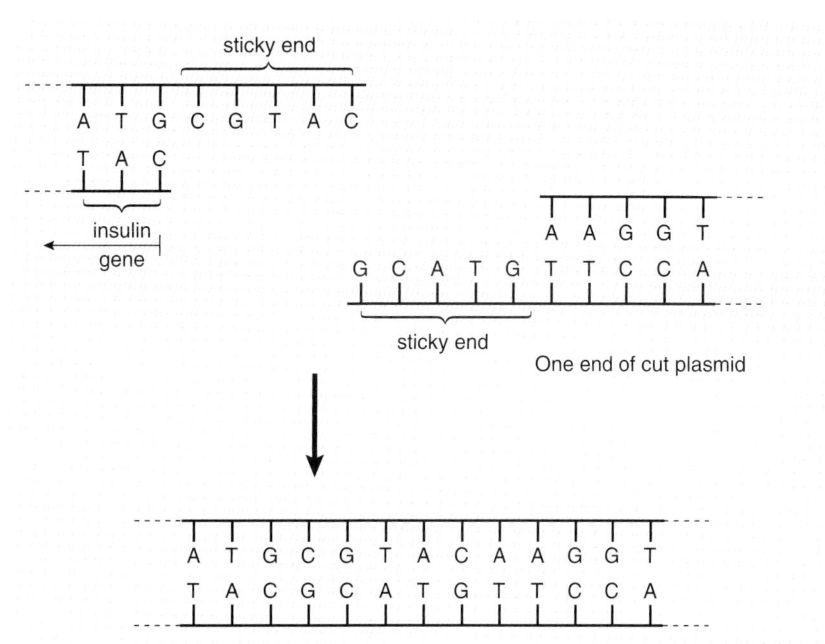

Figure 12.4 Sticky ends allow pairing to take place and links human insulin DNA into the bacterial plasmid

The amount of insulin available was limited by the number of animals brought to the abattoirs for slaughter and it was also a slow extraction process. An additional complication was the fact that non-human insulin differs in structure to human insulin and is therefore not quite as effective for humans. Genetically engineered human insulin overcomes these problems and this method produces the large quantities of insulin that are required today.

Search

▶ genetic engineering and insulin

▶ The need for downstreaming

Following the production of the insulin by genetic engineering (also called recombinant DNA technology) the insulin needs to be **extracted** from the fermenter, **purified** and **packaged** before it can be used for medical purposes. These processes that take place after the insulin is produced by the genetically engineered bacteria are referred to as **downstreaming**.

Question

3 Suggest why downstreaming can be a time-consuming and expensive process.

Many other medical and non-medical products are now made by genetic engineering. In general, pure forms can be produced more quickly, more cheaply and in greater quantities than by older extraction methods.

Exam questions

1 Down syndrome is one example of an inherited disease where all the cells of the individual have 47 chromosomes rather than 46.

The histogram shows how the risk of having a child with Down syndrome increases with the age of the mother.

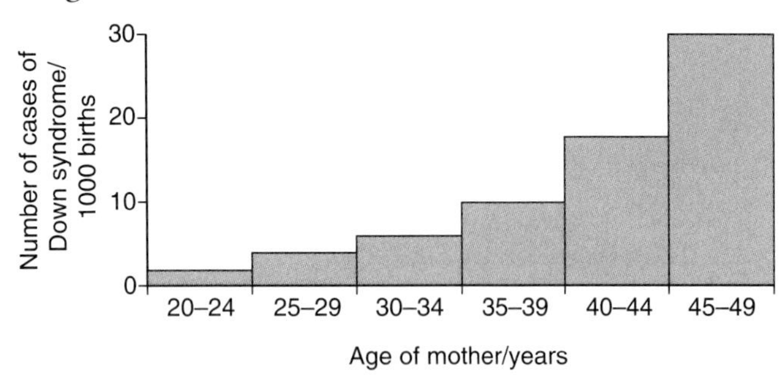

a) Use the histogram to calculate how many times more likely it is for a 42-year-old woman to have a Down syndrome baby than a 32-year-old woman. Show your working. *(2 marks)*

During pregnancy it is possible to detect the presence of Down syndrome in the developing baby by testing a sample of the amniotic fluid.

b) What must be present in the amniotic fluid so that the test can be carried out? *(1 mark)*

2 a) Down Syndrome can be caused by a mutation which results in an extra chromosome.

i) Name the test used to determine whether a foetus has Down syndrome. *(1 mark)*

The diagram shows a foetus in the uterus. To carry out this test, a sample of cells is taken and analysed.

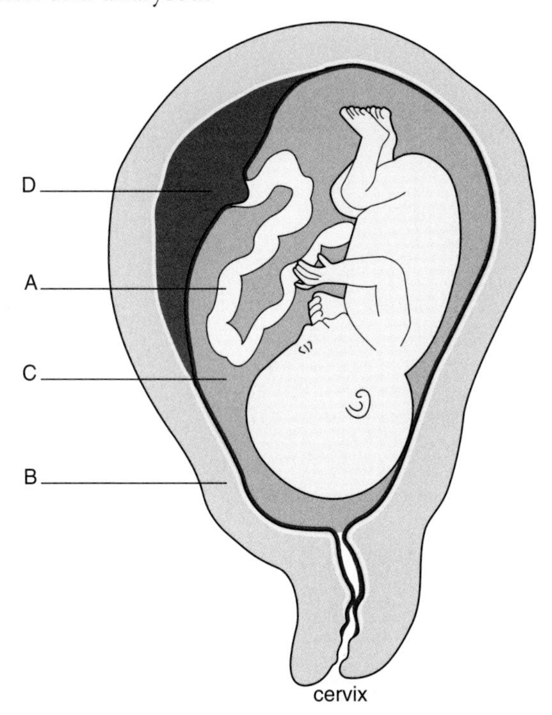

ii) Which letter shows where cells would be collected from to carry out a test for Down syndrome: A, B, C or D? *(1 mark)*

iii) What ethical dilemma could parents face if the test for Down syndrome is positive? *(1 mark)*

b) The table shows how the age of the mother affects the number of babies born with Down syndrome.

Age of mother in years	Risk	Number of babies born with Down syndrome per 1000 births
35–36	1 in 325	3
37–38	1 in 200	
39–40	1 in 110	9
41–42	1 in 77	13

i) Copy and complete the table. Show your working. *(2 marks)*

ii) Describe how the age of the mother affects the risk of having a baby with Down syndrome. *(1 mark)*

iii) The information in the table could be shown in a graph. What data would be plotted on the *x*-axis and *y*-axis? *(2 marks)*

3 a) Use the diagrams and your knowledge to describe and explain how human insulin can be made using genetic engineering.

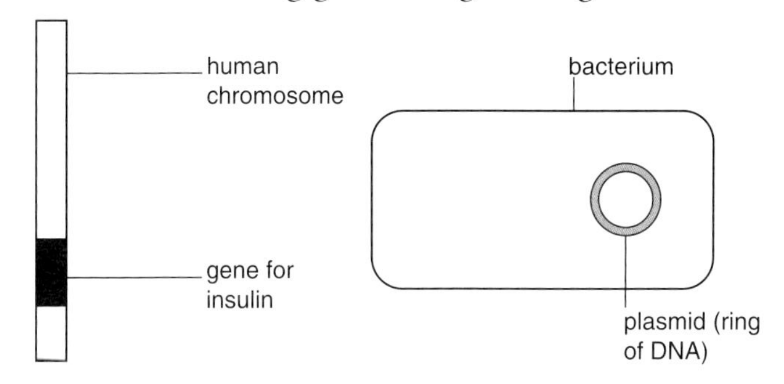

In this question you will be assessed on your written communication skills including the use of specialist science terms. *(6 marks)*

b) Suggest **two** advantages of producing human insulin by this method. *(2 marks)*

13 Variation and Selection

▶ Variation

Living organisms that belong to the same species usually vary from each other in many ways. This variation is caused by differences in **genetic** makeup and/or by **environmental** factors. Differences among individuals for a particular feature are often due to a combination of genetic and environmental factors. A good example is height in humans, which is affected by both genetic and environmental influences. A child's genes will determine the potential height that he or she can reach, but he or she will only grow to that height with a good diet and good overall health. Eye colour, on the other hand, is purely genetic and cannot be affected by environmental conditions. At the other extreme any differences in the appearance of identical twins must be environmental as they have an identical genetic makeup.

Continuous and discontinuous variation

Variation in a particular characteristic can be either continuous or discontinuous. In **continuous variation** there is gradual change in a characteristic across a population. Height is an example of continuous variation in humans. While people can be described as being tall or short, there is not a distinct boundary that separates short and tall people. Figure 13.1 shows a typical set of values for height in human males. Note that the histogram produced shows a normal distribution. A normal distribution is where most individuals are around the average or mean value and relatively few are found at either extreme.

Mass in humans will also show continuous variation.

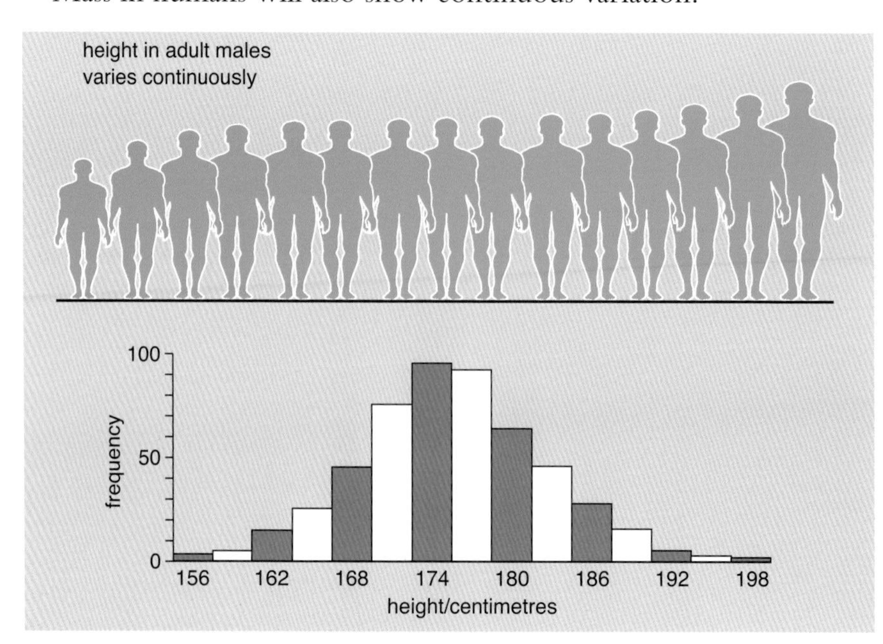

Figure 13.1 Height as an example of continuous variation in humans

Figure 13.2 This child can roll his tongue

In **discontinuous variation** the population can be clearly divided into discrete groups or categories. A common example used is the ability, or inability, to **roll the tongue** in humans. In this example, individuals will fit into one of two categories – there are no intermediates.

In other examples of discontinuous variation there can be more than two categories, for example **ABO blood groups**, but again all individuals can be clearly identified as belonging to a particular blood group (A, B, AB or O).

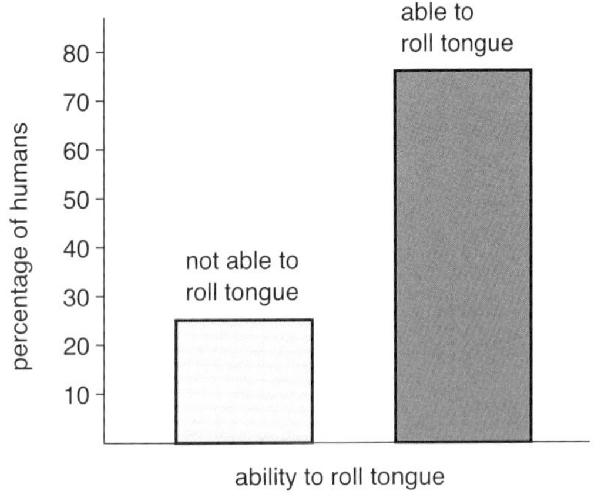

Figure 13.3 Tongue-rolling as an example of discontinuous variation in humans

You should be able to represent information on variation graphically. Continuous variation should be represented as a **histogram** and discontinuous variation as a **bar chart** (with spaces between the bars).

▶ Variation, selection and evolution

If there is a lot of variation among the animals or plants of the same species (type) in an area, it is likely that some of the individuals will be better equipped to prosper or survive in their environment. That is, they are better **adapted**.

Natural selection

In nature, adaptations in living organisms are essential for survival and success in all habitats. It is not difficult to work out some of the main adaptations in polar bears, for example. These adaptations are even more important when organisms compete with each other for resources. This competition ensures that the best-adapted individuals will survive. For example, the larger seedlings growing in a clump of plants will be able to obtain vital resources such as light, nutrients and water more easily than the smaller seedlings. As a result of this competition the stronger individuals will survive, possibly at the expense of the weaker ones. This competition for survival, with the result that the better-equipped individuals survive, is the cornerstone of Charles Darwin's theory of natural selection.

Figure 13.4 Charles Darwin

Charles Darwin and the theory of natural selection

Charles Darwin (1809–1882) was a naturalist who devoted much of his life to scientific research. As part of his research he spent 5 years as a ship's naturalist on the *HMS Beagle* as it travelled to South America. Darwin was greatly influenced by the variety of life he observed on his travels and, in particular, by the animals of the Galapagos Islands. Darwin's famous account of natural selection, *The Origin of Species*, was published in 1859.

Darwin's main conclusions about natural selection can be summarised as:

* There is **variation** among the individuals in a population.
* If there is competition for resources there will be a **struggle for existence**.
* The better-adapted individuals survive this struggle or competition. This leads to **survival of the fittest** and these individuals are more likely to pass their genes on to the next generation.

It is useful to look at an example of natural selection in action to highlight the key features of Darwin's theory.

Antibiotic resistance in bacteria

When bacteria are treated with an antibiotic such as penicillin, most of them are destroyed. However, a small number (*the fittest*) may survive, probably because they have an allele (caused by a mutation) that provides resistance. Very soon the resistant bacteria are the only ones remaining, as they are the only ones surviving and passing their beneficial mutations on to their offspring.

Question

1 Describe and explain how insecticide resistance in insects could arise.

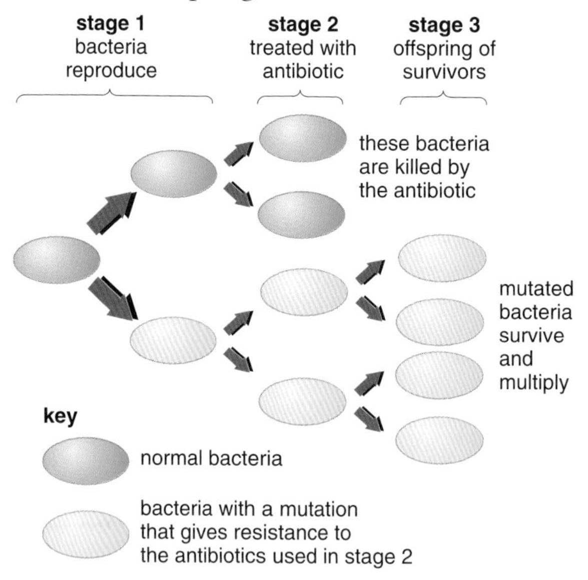

Figure 13.5 Antibiotic resistance in action

The link between natural selection and evolution

Darwin used the theory of natural selection to explain the process of **evolution**. He suggested that *species have changed gradually through time* in response to changes in the environment and that evolution is a *continuing process*. He was not the first to propose that organisms could evolve, but he was the first to propose a plausible theory to explain the mechanism and support it with extensive evidence.

Extinction

An organism may be described as being **extinct** if there are no living examples of it left. The fossil record shows that many animals and plants have become extinct through time. The most famous examples include the dinosaurs and mammoths.

Most of the animals and plants that have become extinct have disappeared because they could not evolve fast enough to cope with a changing environment. The fossil record suggests that in the past, extinctions were often associated with climate change. It is possible that other major events such as meteors striking the Earth could also have disrupted the environment to such an extent that this could have caused the extinction of organisms.

The woolly mammoth (Figure 13.6), a large elephant-like mammal, may have become extinct at the time of the last ice age because the increasing size of the ice fields eliminated much of its natural habitat.

Figure 13.6 Woolly mammoths

Although extinctions have always occurred, there is clear evidence that the activities of humans have been directly or indirectly responsible for the extinction or near-extinction of many plant and animal species. This may result from the direct hunting or collection of animals or plants, but the loss of habitat can have an even more devastating effect. The threat to the panda due to the loss of the bamboo forests and the daily extinctions that probably occur as much of the Amazon rainforest is cleared are well known. However, there are many other local examples where plants and animals have been driven to the verge of extinction or made extinct by the actions of humans e.g. corncrake loss in N. Ireland.

Many more species are **endangered** to the extent that their numbers are so low that extinction is a real possibility in the future. Again, factors such as climate change, the loss of habitats and hunting are all contributing.

Exam questions

1 a) Copy and complete the table.

Type of variation	Human characteristic	Cause of variation
Continuous		Environment
	Tongue rolling	

(3 marks)

b) Explain what is meant by continuous variation. *(1 mark)*

The diagram shows graphs drawn from data collected during biology experiments.

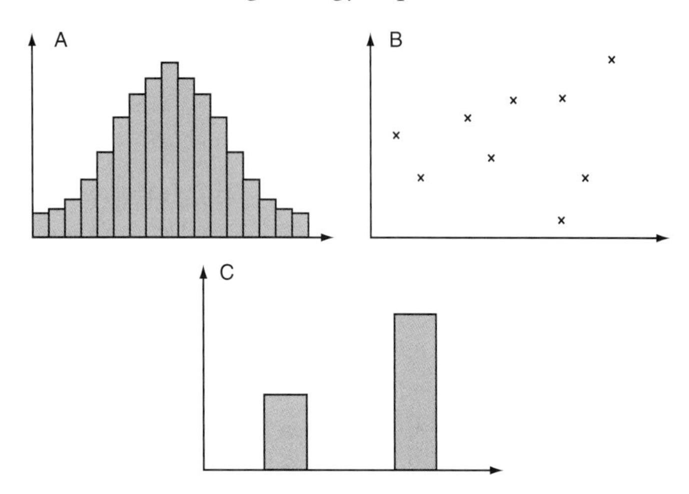

c) Which is the graph for

i) height *(1 mark)*

ii) tongue rolling? *(1 mark)*

2 The illustration shows two forms of the peppered moth, a pale form and a black form. Birds feed on both forms of moth.

a) What would happen to the black moths in a woodland with light-coloured tree trunks? *(1 mark)*

b) A wood had the following proportion of the two forms of moth.

Percentage	
Pale	Black
98	2

The tree trunks became blackened with soot from a nearby factory.
What would happen to the percentage of each type of moth in the wood? *(2 marks)*

3 The following table shows the number of pupils in Year 11 in a school in Northern Ireland who can, and cannot, roll their tongues.

Ability to roll tongue	Number of pupils
Can	50
Cannot	15

a) i) Draw a bar chart using the data in the table. Label each axis. *(3 marks)*

ii) Explain why tongue rolling shows a discontinuous variation. *(1 mark)*

b) The process of evolution is brought about by **natural selection** favouring particular individuals in a population.
An example can be seen in the resistance of some bacteria to antibiotics.
Antibiotic resistance is brought about by a mutation that enables resistant bacteria to survive in the presence of antibiotics.
However, antibiotic-resistant bacteria do not reproduce as well as normal bacteria in an antibiotic-free environment.
The following table shows the relative numbers of normal bacteria and antibiotic-resistant bacteria in two different environments.

Environment	Normal bacteria/%	Antibiotic-resistant bacteria/%
Antibiotic-free	99	1
Antibiotic present	0	100

Explain these results, in terms of natural selection.

In this question, you will be assessed on using your written communication skills including the use of specialist science terms. *(6 marks)*

4 Clover is a type of plant commonly found in lawns. The clover exists in two varieties:

A normal clover that is a common food of slugs
B a slow growing variety that is poisonous to slugs.

The table shows the effect that increasing slug numbers has on the percentage of the two varieties over a period of years.

Year	% Clover in lawn		Number of slugs
	A	B	
2000	80	20	44
2001	75	25	60
2002	62	38	84
2003	50	50	100
2004	42	58	120

a) Name the type of selection shown. *(1 mark)*

b) Explain the change in the types of clover over the period shown. *(3 marks)*

c) State what is meant by the term 'evolution'. *(1 mark)*

14 The Circulatory System

The circulatory system has three main components: the blood; the blood vessels; and the heart that pumps the blood. It has two main functions: **transport** of blood cells, absorbed food molecules (e.g. glucose and amino acids), carbon dioxide, hormones and urea; and **protection** against disease.

Blood

You should examine a blood smear using a microscope. Figure 14.1 shows the main components of blood.

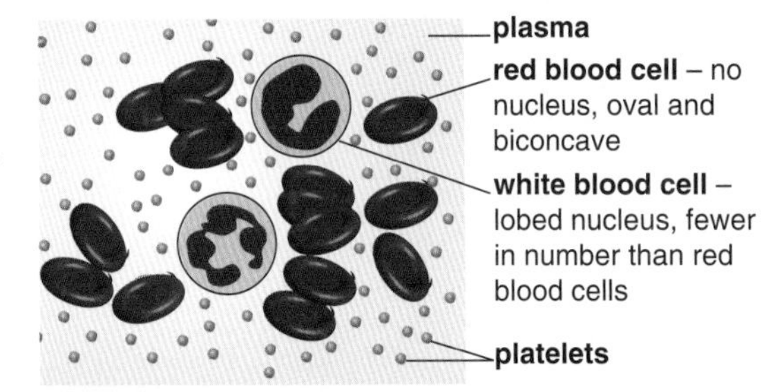

Figure 14.1 The main components of blood

The main components of blood are:

* **Red blood cells** – the function of these cells is to carry **oxygen** around the body.

  It is the **haemoglobin** (rich in **iron**) in the red blood cells that enables them to do this. The **biconcave shape** gives a large surface area for diffusion. The **absence of a nucleus** means that more haemoglobin can be carried in each cell.

* **White blood cells** – the blood contains two types of white blood cell and both are important in defence against disease. **Lymphocytes** produce antibodies and **phagocytes** engulf and digest microorganisms in a process called **phagocytosis**.

* **Platelets** – these very small structures are important in **blood clotting** and the formation of **scabs**. The platelets work by converting the protein fibrinogen to fibrin.

* **Plasma** – this is the liquid part of the blood. The plasma is responsible for the **transport** of the blood cells, absorbed food molecules (e.g. glucose and amino acids), carbon dioxide, hormones and urea.

Question

1 Use the information provided from the blood smear and in the section above to state the difference between red and white blood cells in terms of relative size, relative number and structure.

The blood vessels

There are three main types of blood vessel in the body. If you are doing GCSE Biology separate science they should each be examined using a microscope. The following table shows the main differences between **arteries**, **veins** and **capillaries**.

Vessel	Direction of blood flow	Thickness of wall	Blood pressure	Valves
Artery	Away from the heart	Thick	High	None
Vein	Back to the heart	Relatively thin	Low	Yes
Capillary	Links arteries and veins	One cell thick	Low	None

Note:

1 Arteries carry oxygenated blood (rich in oxygen) to the body organs and veins carry deoxygenated blood (blood with little oxygen present) back to the heart. Figure 14.2 shows that the pulmonary artery and the pulmonary vein are exceptions in that the pulmonary artery carries deoxygenated blood and the pulmonary vein carries oxygenated blood.

2 Diffusion of oxygen, carbon dioxide, dissolved food and urea takes place between the capillaries and the body cells (or vice versa). The one-cell-thick walls are thin enough to be permeable and allow the diffusion to take place.

3 The valves in the veins prevent the backflow of blood. By the time the blood reaches the veins the pressure is low and the valves are necessary to keep the flow in the right direction (**unidirectional**).

4 The walls of the arteries contain **muscle** to give strength and to smooth out the blood flow and **elastic fibres** to allow the arteries to expand as the blood is pulsed through them. Veins have less muscle and few or no elastic fibres.

Question

2 Explain the link between the blood pressure in the arteries and veins and the thickness of the vessel walls.

Question

3 a) Describe the functions of the following blood vessels:
   i) the hepatic portal vein
   ii the coronary arteries.
 b) Describe two differences between the blood in each of the following pairs of vessels:
   i) renal artery and renal vein
   ii) hepatic artery and hepatic vein.

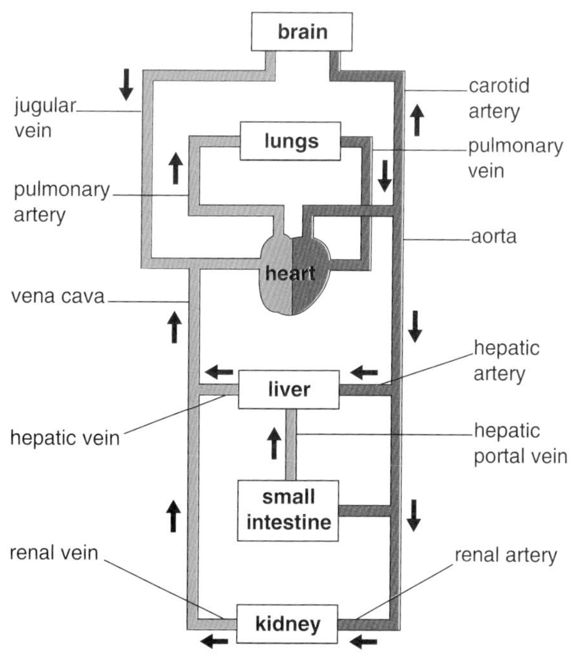

Figure 14.2 The circulatory system

Note: Only students taking GCSE Biology (Higher Tier) need to know the carotid artery and jugular vein.

The heart

The heart is the pump that pumps the blood to the lungs and around the body. This is why the heart has two sides – the right-hand side pumps the blood to the lungs and the left-hand side pumps the blood that has returned from the lungs around the body.

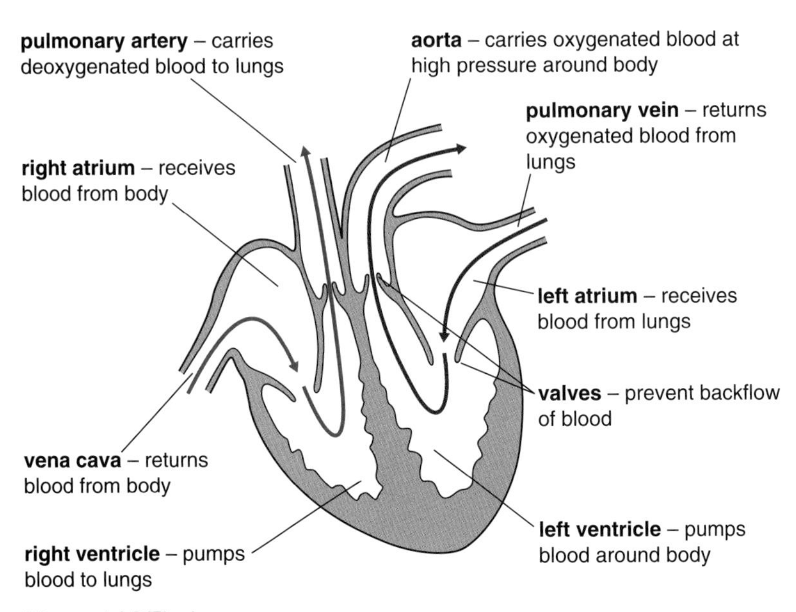

Figure 14.3 The heart

The **right atrium** receives deoxygenated blood from the body. This passes into the **right ventricle** where it is pumped out in the **pulmonary artery** to the lungs. In the lungs the blood becomes oxygenated and returns to the **left atrium** of the heart through the **pulmonary vein**. The oxygen-rich blood passes into the **left ventricle** and is pumped into the **aorta** and then around the body.

The **heart valves** prevent backflow and ensure that the heart acts as a unidirectional pump.

The ventricles are thicker than the atria as they are the chambers that pump blood out of the heart. The left ventricle has a thicker muscular wall than the right ventricle as it pumps blood around the whole body as opposed to just the lungs.

Humans (and other mammals) have a **double circulatory system**. This means that the blood travels through the heart twice in one circulation of the body.

The heart itself receives blood from the **coronary arteries**, which branch from the aorta almost immediately after it leaves the heart. These are the fine blood vessels that can be seen running over the outer surface of the heart.

Search
▶ heart structure

Health and the circulatory system

Heart attacks and strokes

Heart attacks and strokes have been discussed in an earlier chapter on nutrition. However, it is important to remember that they are diseases of the circulatory system.

Heart disease

Heart disease is caused by cholesterol and other fatty substances being present at such high levels that they build up in the walls of the arteries. Over time this leads to a narrowing of the arteries, making it more difficult for blood to flow through them. This is particularly likely to happen in the **coronary arteries** that supply the heart, hence the term **coronary heart disease** (CHD).

Eventually an artery may become so narrow that a blockage forms and stops the blood from flowing in this particular artery. This prevents the heart muscle that the artery serves from receiving **oxygen** and **glucose**. **Respiration** in the heart cells can no longer happen and they die, causing the heart to stop beating. This is a **heart attack**.

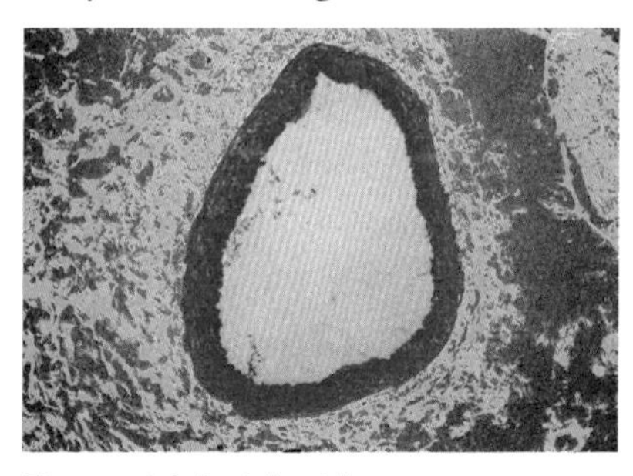

Figure 14.4a A healthy coronary artery

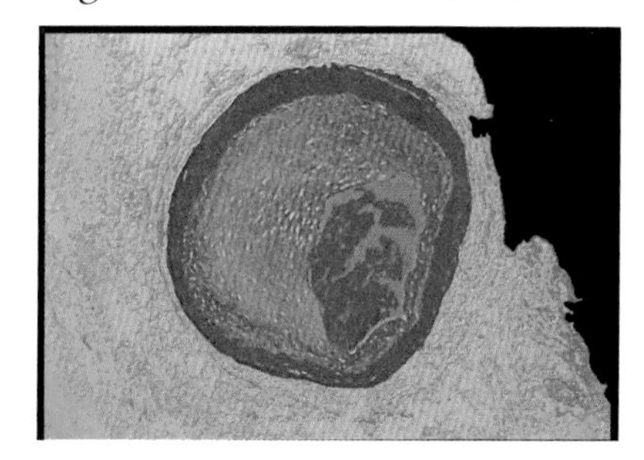

Figure 14.4b A coronary artery that has become blocked by cholesterol (orange)

Strokes

If the blockage is in the brain, a **stroke** can result. Again, the cells deprived of oxygen and glucose die and the affected part of the brain stops functioning properly. This often causes paralysis of parts of the body.

The factors that increase the risk of circulatory disorders are summarised in Figure 14.5.

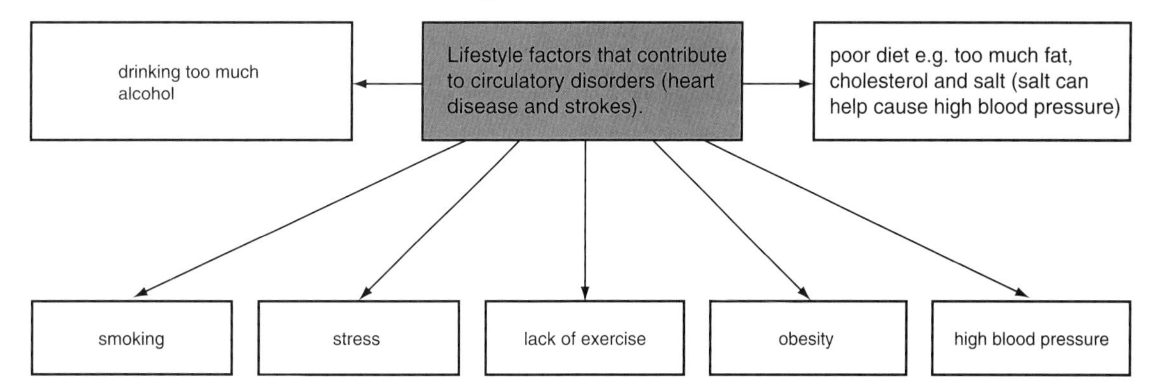

Figure 14.5 Factors that increase the risk of circulatory disorders

As well as **lifestyle factors** (factors that we can do something about), **genetic factors** can also play a part.

Other circulatory diseases

Other circulatory diseases can be caused by an **imbalance** occurring in the level of the different blood components. An example is anaemia, which is a shortage of red blood cells. This can be caused by the body not getting enough iron in the diet or through a loss of blood. The main symptoms of anaemia are:

* extreme tiredness (which can lead to fainting)
* a pale complexion
* a rapid pulse rate (and palpitations).

Question

4 Suggest why the following are symptoms of anaemia:
  a) extreme tiredness
  b) a rapid pulse rate

Exercise and the circulatory system

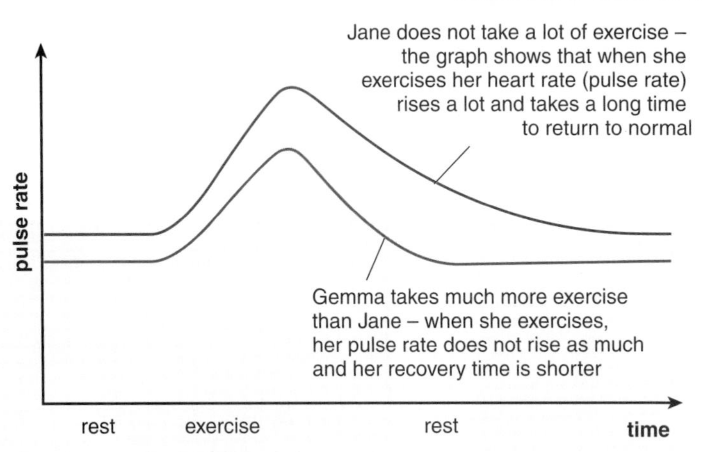

Figure 14.6 The effect of exercise on heart rate

Regular **exercise** benefits the circulatory system in a number of ways. As already noted it can help reduce the risk of heart disease or strokes by burning up fat that could otherwise clog up the arteries or lead to obesity. Exercise also helps by strengthening the heart muscle (as with any muscle that is exercised). A stronger heart will have an increased output (pumps more blood per beat) even when not exercising. This means that the heart is under less strain as it has to pump less often to get the same amount of blood around the body over a period of time.

You should also investigate the effect of exercise on **pulse rate**. The graph in Figure 14.6 shows the effect of exercise on heart rate and recovery rate.

* The recovery rate is the time it takes for the pulse or heart rate to return to normal after exercise and this will usually be shorter for people who exercise or play a lot of sport.
* You should also know why exercise causes the heart rate to rise. When we exercise we need more energy and the heart has to pump more blood to our muscles so that they get more oxygen for respiration.

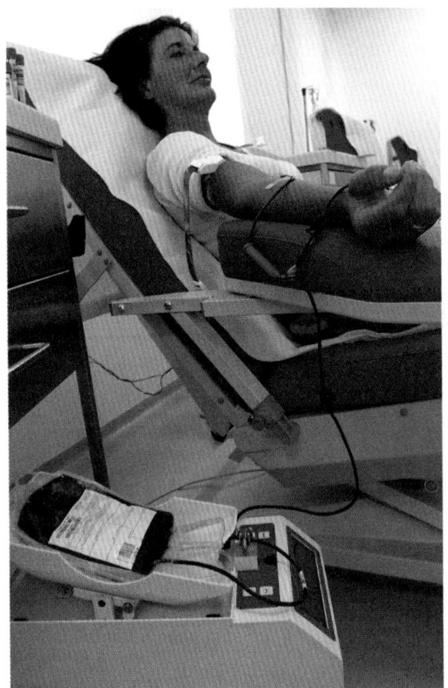

Figure 14.7 Donating blood

Blood donation

Blood donors provide blood (**blood donation**) on a voluntary basis and this blood is used for medical purposes. Some **blood disorders** can be treated by giving a patient a specific component of the blood if they are suffering from a shortage of that component. For example, some patients are given **transfusions** of platelets or specific clotting agents from the blood. Many other people need to have blood transfusions following accidents or surgery. These transfusions often contain mainly red blood cells.

Blood donors in Northern Ireland

People aged 17–70 who have 'healthy blood' can give blood in Northern Ireland. However, only about 7% of those people who are eligible to donate blood do so regularly. The small number of regular donors, together with the increasing demand for blood for transfusions means that there is a shortage of donors in Northern Ireland. A further factor is that the donated blood can only be stored for a limited period of time. Some religious groups do not believe in donating blood or receiving blood transfusions when ill.

The Northern Ireland Blood Transfusion Service (www.nibts.org) is a good place to find more information.

Search
▶ Northern Ireland Blood Transfusion Service

▶ Plasma, tissue fluid and lymph

When blood reaches the capillaries, the liquid part (**plasma**) passes through the thin walls and bathes the body tissues (cells) due to the pressure of the blood flow. Plasma is rich in glucose and oxygen (in the capillaries the haemoglobin releases its oxygen). The cells of the body are surrounded by a liquid called **tissue fluid** and the plasma becomes part of the tissue fluid. Most of the liquid that leaves the capillary eventually diffuses back into it, carrying carbon dioxide and any other wastes produced by the body cells.

However, some of the tissue fluid ends up too far away from the capillary to pass back into it. This fluid is then 'mopped up' by lymphatic vessels (similar to the lacteals you first came across in Chapter 4 on the digestive system). The fluid is now called **lymph** and is eventually returned to the blood.

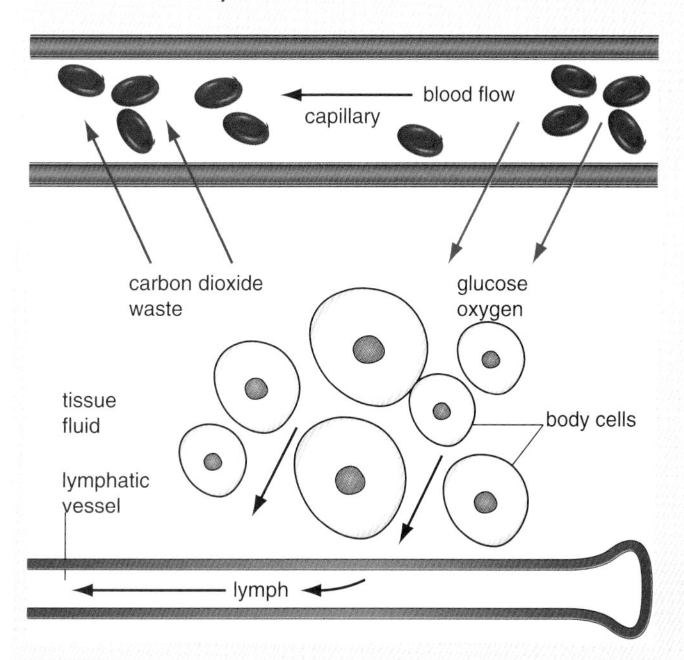

Figure 14.8 The formation of tissue fluid and lymph

Plasma, tissue fluid and lymph are fairly similar in composition. The main difference is in their location.

Exam questions

1 a) Give **two** functions of the circulatory system. *(2 marks)*

b) An outline of the human circulatory system is shown.

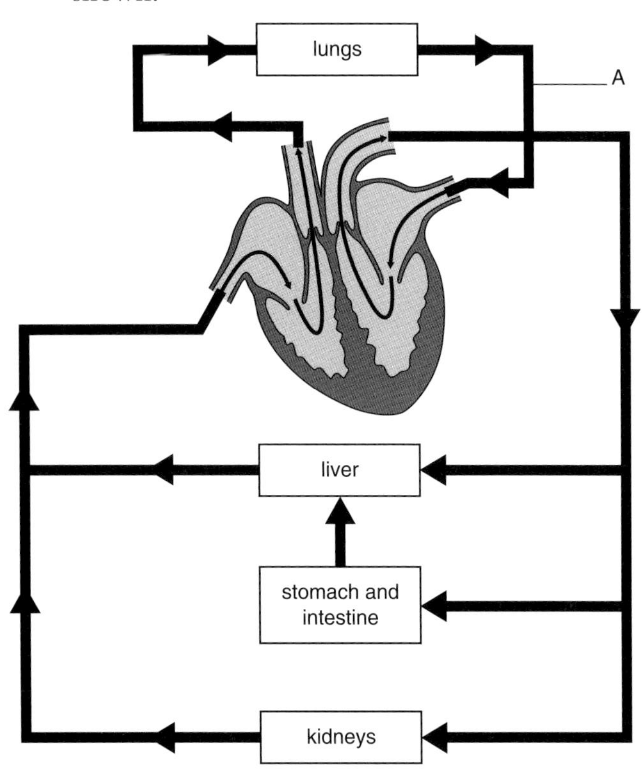

i) Name blood vessel A. *(1 mark)*

ii) Give **one** difference in the composition of the blood in vessel A and in the vena cava. *(1 mark)*

c) Name the blood vessel, not shown in the diagram, that can become blocked resulting in a heart attack. *(1 mark)*

2 The diagram shows an external view of the heart.

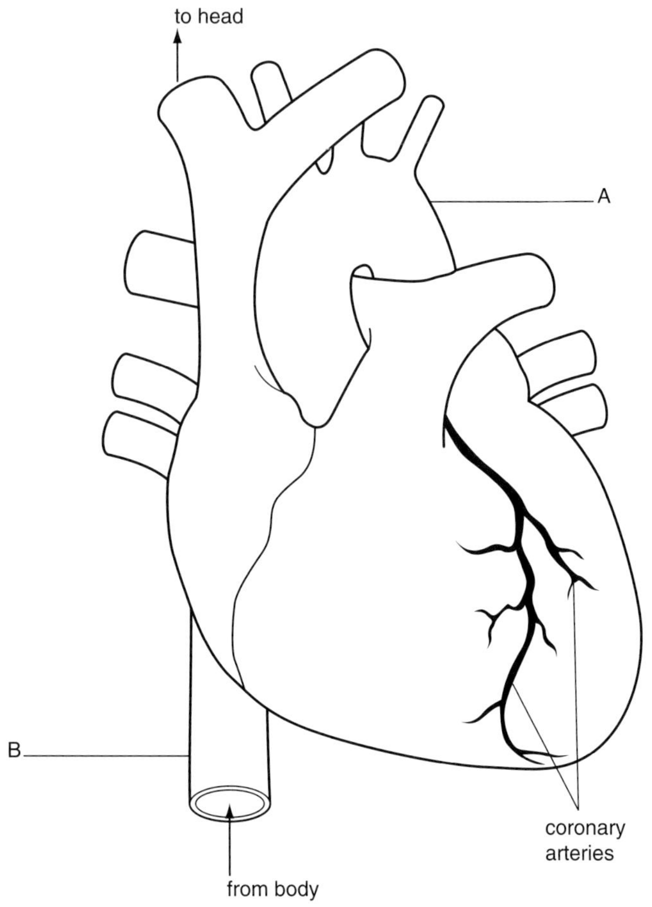

a) Name the blood vessels A and B. *(2 marks)*

The diagram shows a section through a normal coronary artery and one from a heart attack victim.

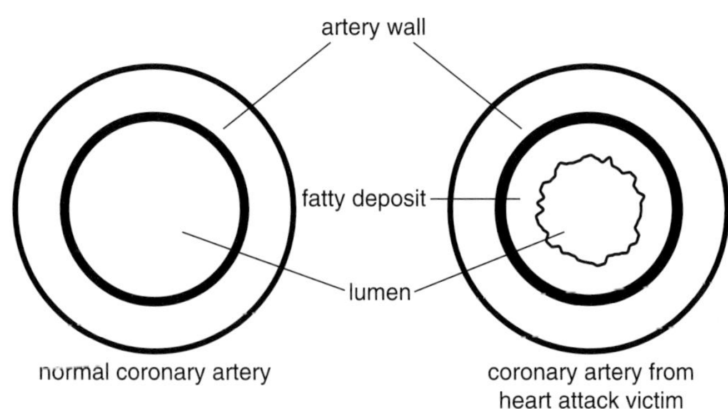

b) Use the diagram to suggest how a fatty deposit on the wall of the coronary artery causes a heart attack. *(3 marks)*

c) Give **two** changes in lifestyle that could reduce the risk of a further heart attack. *(2 marks)*

3 The diagram shows a CT scan from a person who has suffered a stroke.

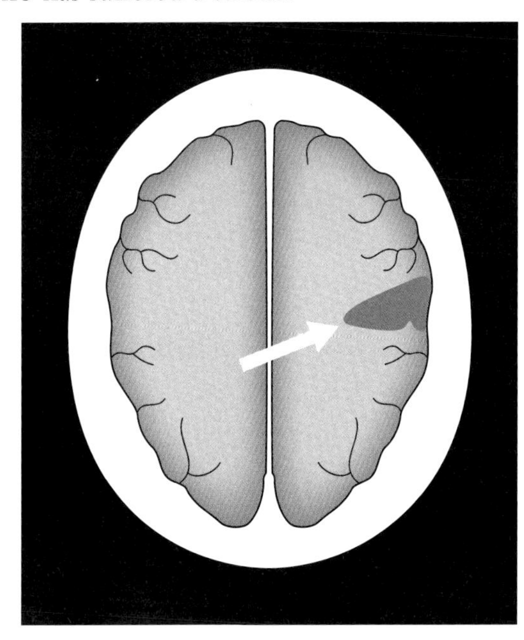

a) i) Different parts of the brain control different functions of the body. Suggest how the CT scan links with the fact that one side of the person's body has been affected by this stroke. *(1 mark)*

ii) Suggest how the size or location of the areas of the brain affected may be linked with the symptoms the person shows following the stroke. *(1 mark)*

iii) Explain how a stroke occurs.
In this question, you will be assessed on using your written communication skills including the use of specialist science terms. *(6 marks)*

b) Some people are more likely than others to suffer a stroke. Nine out of ten people affected are over 55. Certain risk factors increase the chances of someone having a stroke including genetic factors, high blood pressure and diabetes.

i) Give **two** risk factors linked to strokes that are out of a person's control. *(2 marks)*

ii) Give **three** pieces of advice that could lead to a reduction in the chances of someone developing a stroke.
Use the information from above and your knowledge to help answer the question. *(3 marks)*

c) i) Blood is transported around the body by a series of blood vessels.
Name the blood vessels that carry blood away from the heart. *(1 mark)*

The following diagram shows a capillary in cross-section.

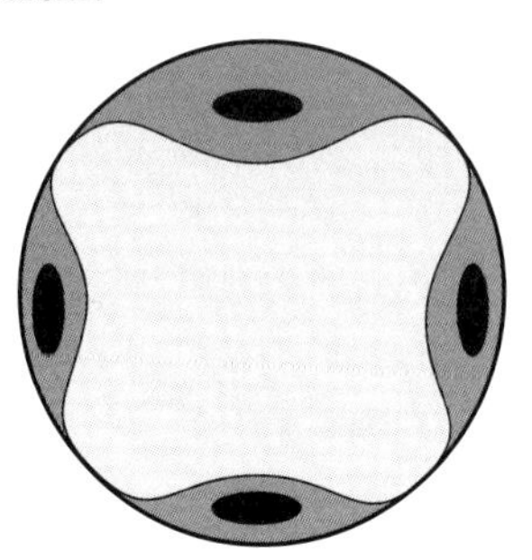

ii) Use the diagram and your knowledge to suggest how its function is related to its structure. *(1 mark)*

4 An investigation was carried out to compare the recovery of pupils of different athletic ability after a period of strenuous exercise.

Two athletic and two non-athletic pupils ran 200 metres. Their pulses were counted over the 2 minutes immediately after the exercise.

Pulses were recorded for each of four consecutive 30 second periods.

The results for each pupil are given below:

Athlete A: 75, 55, 40, 30

Athlete B: 78, 53, 42, 28

Non-athletic C: 71, 64, 58, 39

Non-athletic D: 67, 58, 46, 41

a) Organise these results into an appropriate table.
Your table should have suitable column headings and include results. *(3 marks)*

b) Describe **two** trends in the results. *(2 marks)*

c) i) Which group had the greatest decrease in pulse in the first minute after exercise? *(1 mark)*

ii) Suggest why this group had the greatest decrease in pulse in the first minute after exercise. *(1 mark)*

5 A blood donation occurs when a healthy person voluntarily has blood taken and used for transfusion.

In Northern Ireland over 62 000 blood donations are given each year from approximately 65 000 donors. Anyone between the ages of 17 and 65 can give blood for the first time, provided there is nothing in their medical history that would endanger their own health or might make their blood unsafe for the recipient.

Donors can continue to give blood up to three times a year until the age of 70.

Around 500 patients each week in Northern Ireland need blood. To meet supply needs, the transfusion service tries to keep enough blood for 5 days in stock. To maintain this stock the service needs to recruit about 9 000 new donors each year.

a) Explain why a person with a medical history of anaemia would not be accepted as a blood donor. *(2 marks)*

Give **two** reasons why a patient may need blood. *(2 marks)*

Suggest **two** reasons why there is a shortage of blood donors. *(2 marks)*

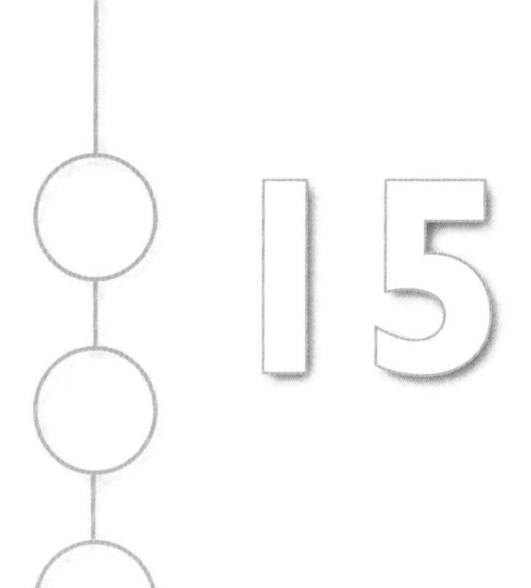

15 Microorganisms, Defence against Disease, Medicines and Drugs

It is very easy to take good health for granted, but from time to time we all fall ill. Most illnesses in this part of the world, while often making us feel very poorly, usually only last for a short period of time.

These illnesses are often caused by very small living organisms called **microorganisms** or microbes that need to gain entry to the body before causing it harm. Microbes are so small that they can usually only be seen using powerful microscopes. There are three main types of microbe that can cause infection: **bacteria**, **viruses** and **fungi**. Microbes that cause disease are referred to as pathogens.

Louis Pasteur, a very famous scientist, carried out a well-known experiment to show the role of microbes in the contamination of drink or food.

Pasteur's work

Before Pasteur's famous work it was assumed that when wine, juice or milk became contaminated, the microorganisms causing the contamination spontaneously appeared (they appeared out of nowhere!). This was the theory of **spontaneous generation**.

Pasteur carried out his research using strangely-shaped 'swan neck' flasks to see if the microbes did suddenly appear from nowhere or if they in fact came from the air.

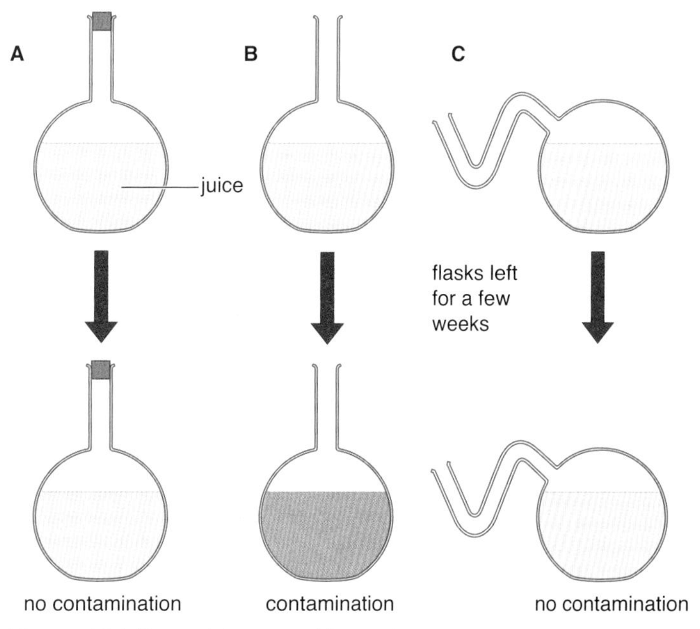

Figure 15.1 Pasteur's 'swan neck' experiment

Pasteur proved that microorganisms were not spontaneously created. Only the juice in Flask B became contaminated, where microbes could gain entry through the open neck. Pasteur's work has helped us understand a feature of many well-known diseases in humans – the microbes have to gain entry to the body before they can cause harm.

Questions

1 a) In Pasteur's experiment, why would the flasks and their contents need to be sterilised before being left?

b) Explain why the juice in flasks A and C did not get contaminated.

Search ▶ pasteurisation and milk

Question

2 Give one reason why pasteurisation does not involve heating the milk to boiling point.

Knowing that contamination of food is caused by microorganisms has led to the development of **pasteurisation** and other techniques that kill or stop the growth of most microorganisms in foods such as milk.

Pasteurisation techniques generally involve heating the milk or other product to a high temperature (but below the boiling point) for a specified short period of time, followed by rapid chilling to a low temperature. This kills most bacteria present.

▶ Some examples of diseases caused by microbes

Bacterial diseases include **tuberculosis** (which mainly affects the lungs), **salmonella** (which is an important cause of food poisoning) and the sexually transmitted diseases **gonorrhoea** and **chlamydia**. **Viral diseases** include **AIDS**, **mumps**, **measles**, **rubella**, **polio** and **colds** and **flu**. A relatively common disease caused by a **fungus** is **athlete's foot**.

Most of these diseases are infectious diseases, that is, they are passed from one person to another. The following table summarises the methods of spread and treatment for these microbes.

Microbe	Type	Spread	Control/prevention/treatment
HIV which leads to AIDS	Virus	Exchange of body fluids during sex Infected blood	Using a condom will reduce risk of infection, as will drug addicts not sharing needles No cure
Rubella	Virus	Airborne (droplet infection) through coughing and sneezing	Prevented by MMR vaccination
Measles	Virus	Airborne (droplet infection) or by contact	Prevented by MMR vaccination
Mumps	Virus	Airborne (droplet infection)	Prevented by MMR vaccination
Colds and flu	Virus	Airborne (droplet infection)	Flu vaccination for targeted groups
Polio	Virus	Usually spread through drinking water contaminated with faeces	The polio vaccination has currently eradicated polio in the UK
Salmonella food poisoning	Bacterium	From contaminated food	Always cooking food thoroughly and not mixing cooked and uncooked foods can control spread Treatment by antibiotics
Gonorrhoea	Bacterium	Sexual contact	Using a condom will reduce risk of infection Treatment by antibiotics
Tuberculosis	Bacterium	Airborne (droplet infection)	BCG vaccination If contracted, treated with drugs including antibiotics
Chlamydia	Bacterium	Sexual contact	Using a condom will reduce risk of infection Treatment by antibiotics
Athlete's foot	Fungus	Contact	Reduce infection risk by avoiding direct contact in areas where spores are likely to be present, e.g. wear 'flip flops' in changing rooms/swimming pools

How can we defend against infectious disease?

The human body is well adapted to protect us against infection. The body is very successful at preventing most microorganisms from gaining entry and it has very effective defences against those microorganisms that do enter.

The **skin** itself is an excellent barrier to microorganisms. The openings in the body such as the nose and the respiratory system are protected by **mucous membranes** that trap the microorganisms and prevent them going any further. **Clotting** is also important as a defence mechanism. It stops more blood escaping but it also acts as a barrier against infection.

If a microorganism does enter the body it is the blood system that usually helps combat the invader. The blood system is very effective in this role but we are often ill for a period of time before our defence system gains the upper hand.

Antigens and antibodies

Invading microorganisms have chemicals on their surface that the body can recognise as being foreign. These chemicals are called **antigens** and they cause special white blood cells called **lymphocytes** to produce **antibodies**.

As Figure 15.2 shows, these antibodies have a shape that matches the shape of the antigens on the microorganisms. The antibodies join with the microorganisms (like a jigsaw puzzle) and cause them to clump together. Once clumped or immobilised they are easily destroyed by other white blood cells called **phagocytes** in a process known as **phagocytosis**.

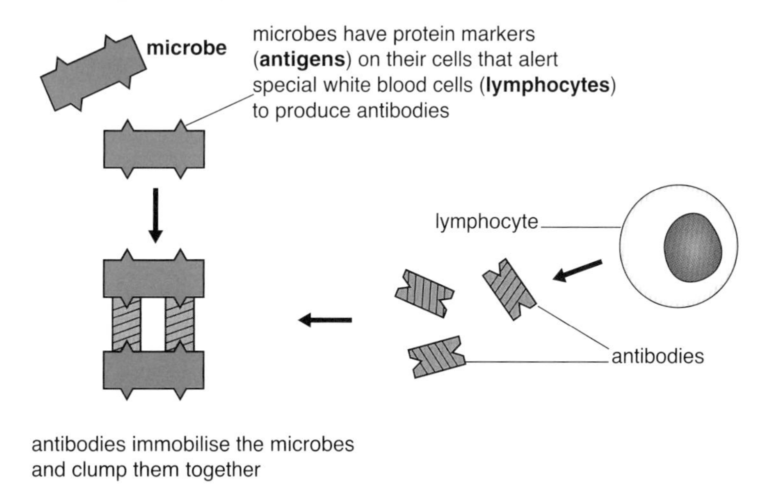

Figure 15.2 How antibodies work

Phagocytosis in action

Some white blood cells move around in the blood and destroy microorganisms trapped by antibodies, or they can destroy them directly without antibody action. The white blood cell (or phagocyte) surrounds the microorganism and **engulfs** ('eats') it, as seen in Figure 15.3. Eventually chemicals (enzymes) inside the phagocyte **digest** the microorganism and destroy it.

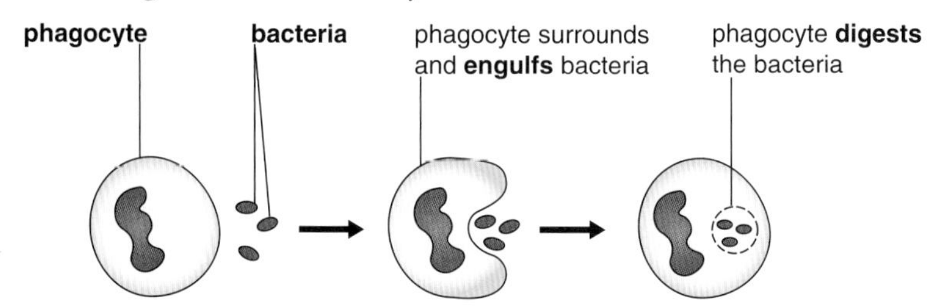

Figure 15.03 Phagocytosis

Immunity

Individuals who are protected against a particular infection or disease are described as being **immune** to that disease. Most people will be immune to a number of diseases. If someone is immune this means that his or her antibody levels are high enough (or high enough levels can be produced quickly enough) to combat the microorganism should it gain entry to the body again. Figure 15.4 shows what happens in the period following infection. The levels of antibody rise as the white blood cells that produce the correct antibody are stimulated by the presence of the microorganism's antigens. The graph shows that after a period of time the antibodies reach a level that is:

* high enough to combat the infection and allow the patient to recover
* high enough to prevent the individual becoming ill again from that particular disease in the future. (In reality it is not just that the antibodies remain at a high enough level in the blood, but that the first infection causes the production of special **memory lymphocyte** cells. These can very rapidly produce the antibodies that match the antigens on this microorganism again should a re-infection occur – known as the **secondary response**.)

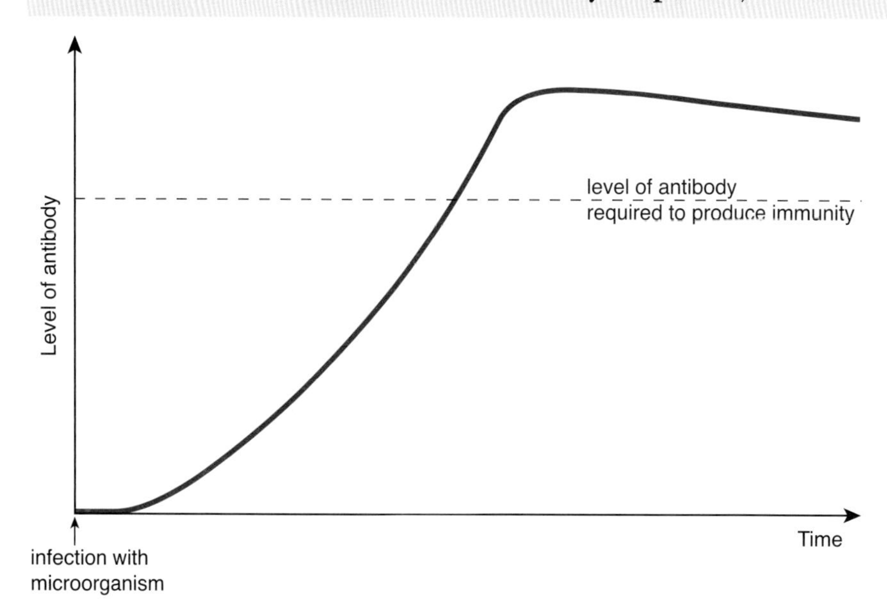

Figure 15.4 The effect of infection on antibody level

This type of immunity is called **active immunity** as it is *the body that produces the antibodies* to combat the invading microorganism. Figure 15.4 shows that it takes time to reach the required antibody level in active immunity and this can be a problem in some situations, such as when an individual has been infected by a serious disease-causing microorganism for the first time.

Passive immunity can be used when rapid protection is required. Passive immunity is the use of ready-made antibodies that are injected into the body. Obviously they can act *very rapidly* but they only last for a *short time* as there are no memory cells to produce more antibodies. Figure 15.5 shows the effect of passive immunity on antibody levels.

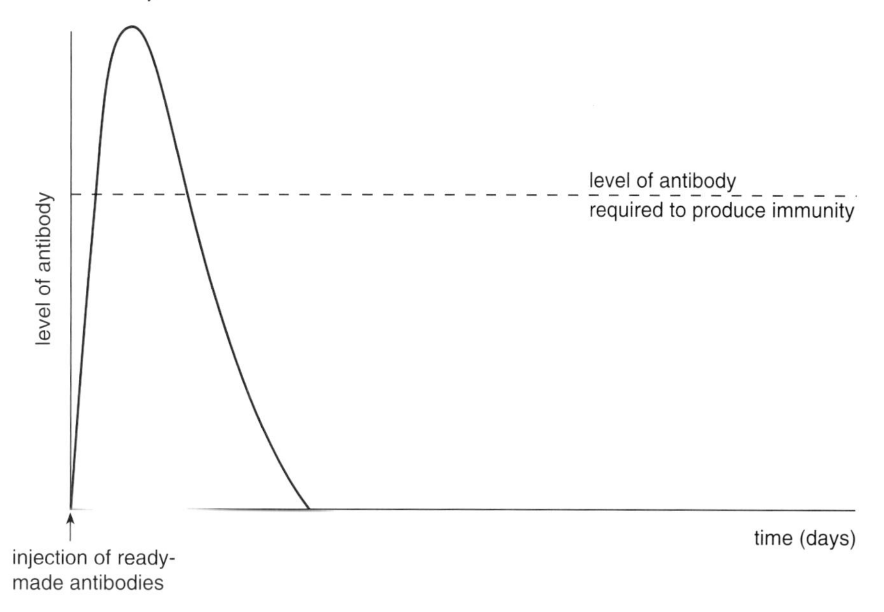

Figure 15.5 The effect of passive immunity on antibody level

Active immunity can also be developed through the use of **vaccinations**.

Vaccinations

Vaccinations involve the use of **dead** or **modified** pathogens (disease-causing microorganisms) that are injected into the body. The dead or modified microorganisms still have the antigens on their surfaces that cause the body to produce antibodies (and memory lymphocyte cells) at a high enough level to prevent the individual becoming ill later. The process is exactly the same as if you had caught the disease – the big difference is you don't get ill first! Sometimes we need more than one vaccination to make sure that we remain immune for a reasonable period of time. This is known as a follow-up **booster**. Figure 15.6 shows what happens following a vaccination that involves a booster.

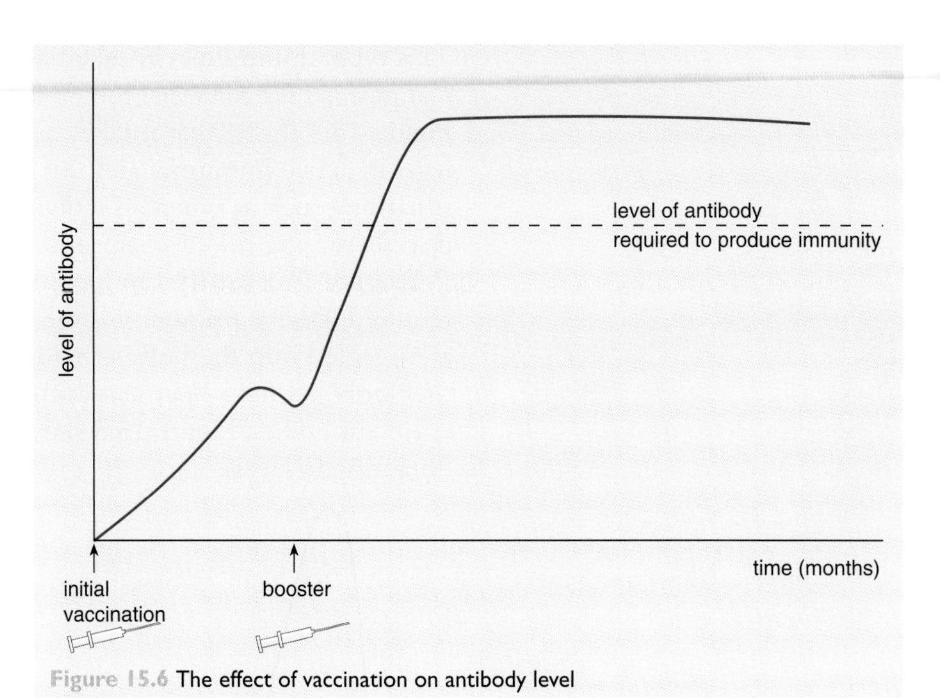

Figure 15.6 The effect of vaccination on antibody level

Types of immunity

We have already looked at passive and active immunity. However, immunity can be further subdivided as shown in Figure 15.7.

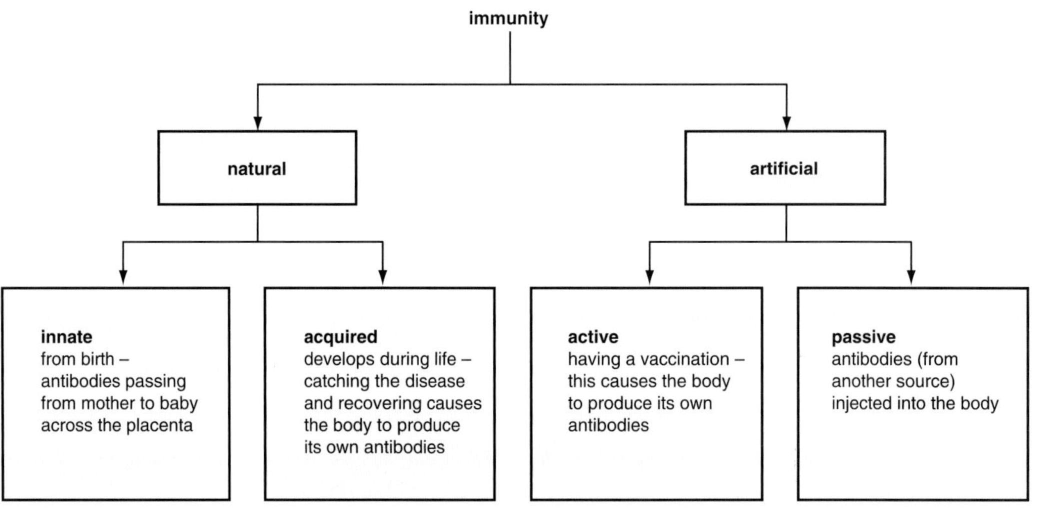

Figure 15.7 The classification of immunity

The first vaccination – the work of Edward Jenner

Jenner noticed that milkmaids who had had the minor illness cowpox did not then catch the serious illness smallpox. He concluded that having had the cowpox protected them from smallpox. In 1796 he tested this by deliberately infecting a young boy (James Phipps) with cowpox by inserting some pus from a cowpox blister into a cut in James' arm. After a period of time Jenner obtained some pus from someone with smallpox and inserted this into James' arm too. James did not catch smallpox – the cowpox microorganism was so similar to the smallpox microorganism that he had built up immunity to both. This was the first vaccination. Clearly, Jenner was taking a big risk in carrying out this experiment!

Search
▶ Jenner and vaccinations

Questions

3 Why was it important that Jenner waited 'a period of time' before infecting James with the smallpox pus?

4 Why do you think most immunisation programmes are targeted at young children and not adults?

5 Why is it important that the microorganisms used in vaccination are dead or modified in some way?

6 Summarise the main differences between passive and active immunity. What are the main advantages and disadvantages of each?

Question

7 Vaccinations that are necessary for foreign holidays often have to be paid for as they are not provided free by the NHS. Can you suggest why most vaccination programmes (e.g. MMR) are free but vaccinations associated with foreign travel are not?

Travel and vaccination

People who travel to some foreign countries are advised or required to be vaccinated against certain diseases. This is not necessary for all countries. For example, we can travel to France or Spain without needing an immunisation programme first, but for much of Africa and the Far East it is a different matter. This is because some countries have pathogens that cause serious illnesses that do not exist (or are not common) in the British Isles. We will not have built up any antibody defences against these pathogens as we have not been in contact with their antigens before. It is not difficult to imagine the harm that could be caused by arriving back home from a holiday with an infectious disease that is usually not found in Britain.

Working with microorganisms in the laboratory – the use of aseptic techniques

In a school environment there are important health and safety precautions that need to be used when growing or culturing microbes. These include:

- not eating or drinking in the laboratory
- using disposable gloves and laboratory coats
- wiping down lab benches with a disinfectant
- culturing and examining microbes in sealed containers
- not culturing microbes at body temperature
- using sterile loops for transferring cultures
- flaming the necks of culture bottles to prevent contamination
- sterilising or disposing of all equipment after use
- washing hands thoroughly after each part of the experiment.

It is very important that the microbes you are working with do not contaminate anything else, and the safety measures described above will help prevent this. It is also important that the microbes themselves are not contaminated by other microbes in the air or on surrounding surfaces. The use in the laboratory of the **aseptic techniques** described above helps to prevent contamination.

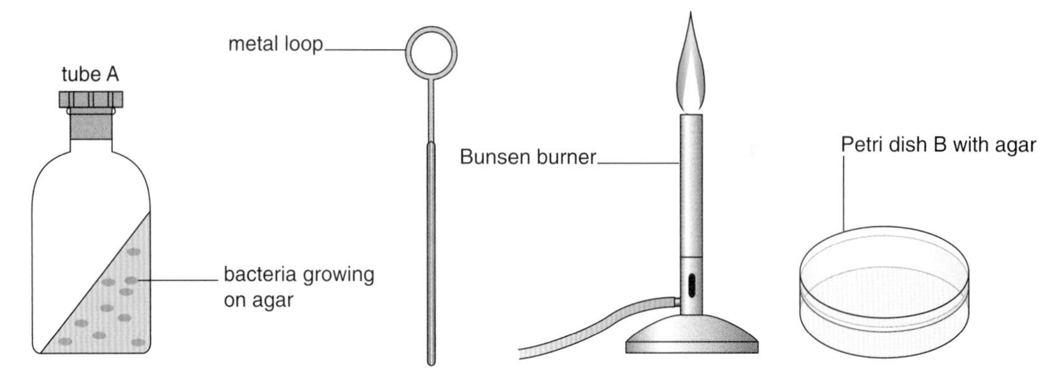

Figure 15.8 Using aseptic techniques

The apparatus in Figure 15.8 is normally used when inoculating and plating bacteria in the laboratory. A typical sequence is:

1 Heat the metal loop by passing it through the flame of the Bunsen burner to sterilise it.
2 Allow the metal loop to cool, as if it is too hot it will kill any microbes that it comes into contact with.
3 Remove the lid of the culture bottle (Tube A) and scrape the metal loop gently over the agar. This will ensure that the metal loop has bacteria from the culture bottle over its surface (inoculation).
4 Replace the lid of the culture bottle to prevent contamination. It is advantageous to 'sweep' the neck of the bottle through the flame to destroy any airborne microbes.
5 Spread the microbes over the surface of the agar in the Petri dish (B) by gently scraping the metal loop over the agar surface (plating). (The agar contains nutrients that help the microbes to grow.) It is useful to hold the Petri dish lid at an angle rather than completely removing it, as this will reduce the chance of unwanted microbes from the air entering the dish.
6 The metal loop can then be heated again to a high temperature to ensure that any microbes remaining on the loop are destroyed.
7 The Petri dish should be sealed with tape and then incubated in an oven at 25 °C. This temperature will allow the microbes to grow and form colonies but is below body temperature, meaning pathogenic microbes that could harm humans will not grow.

Note: instead of using a metal loop it is possible to use sterile disposable plastic loops that do not require heating.

Antibiotics

Chemicals can damage and kill living cells. Some chemicals are produced with the specific effect of destroying harmful microbes. **Antibiotics** are an example.

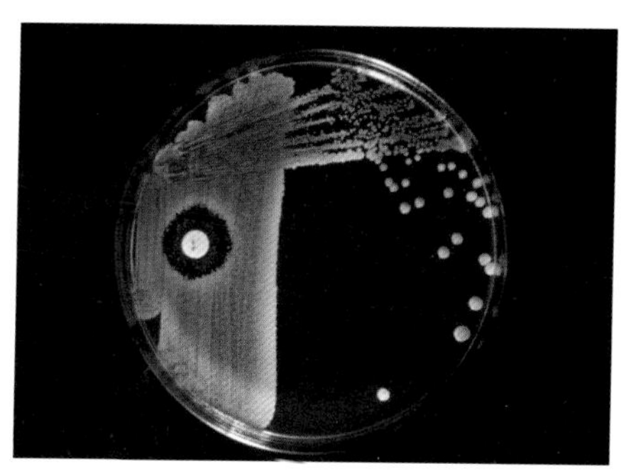

Figure 15.9 Agar plate showing a clear area (around the white circle on the left) where the antibiotic (penicillin) has killed bacteria growing on the agar

Antibiotics are chemicals that can be used to combat bacterial infections. Most people have had antibiotics at some time in their lives to defend against bacterial conditions such as septic throats or infected wounds in the skin. The antibiotics kill bacteria or stop their growth. The effect of an antibiotic can be seen in Figure 15.9.

Antibiotics are not as specific as antibodies in that they are not designed to combat only one type of bacteria – they usually act against a range of bacteria and they act in a different way to antibodies. However, different types of antibiotics have different effects against different bacteria. For this reason a GP may prescribe different antibiotics at different times for the same patient if the bacterial infections are different.

Bacterial resistance to antibiotics

Bacterial resistance to antibiotics was used as an example of natural selection in Chapter 13. Bacterial resistance to antibiotics is becoming a major problem and is making many antibiotics ineffective against various bacteria. The overuse of antibiotics is largely responsible and it is very important that antibiotics are only used when they are really necessary. The overuse of antibiotics has allowed many types of bacteria to become resistant to the main antibiotics, and so it is the mutated resistant forms that are now common.

Questions

8 Siobhan and Catherine are sisters. Siobhan had a bad cold and Catherine had a throat infection caused by bacteria. Their mother took them to the doctor, who gave Catherine an antibiotic but gave Siobhan nothing.

a) Explain why the doctor:
   i) treated Catherine with an antibiotic
   ii) didn't give Siobhan an antibiotic.

b) Suggest why new types of antibiotic need to be continually produced.

c) Suggest why doctors ask patients to make sure they complete a course of antibiotics and do not stop taking the prescribed dose even if they feel better.

'Superbugs' and MRSA

Figure 15.10

Some bacteria have developed resistance to the extent that they are now referred to as **'superbugs'**. They are responsible for a number of serious medical conditions. These superbugs, such as **MRSA**, are resistant to most types of antibiotic and can be a very serious problem in hospitals. The headline in Figure 15.10 is typical of many headlines that have appeared in the press lately.

There has been media speculation that the problem with superbugs is due to poor standards of hygiene in some hospitals, but is this really fair? There is no doubt that good hygiene is very important in preventing the spread of microbes in hospitals but other factors are important in allowing superbugs to flourish in this type of environment too. Patients in hospital often have weak immune systems and they may have wounds that allow microbes to enter the body. Another factor is that hospitals provide an 'antibiotic-rich' environment where the microbes have every opportunity to come into contact with a range of antibiotics. This ensures that the non-resistant microbes are eliminated and a high proportion of the surviving microbes are antibiotic resistant.

There is no doubt that 'superbugs' are extremely difficult to eradicate. Nonetheless, new measures in hospitals include increased levels of hygiene (such as the immediate cleaning of spillages of body fluids and the wearing of gloves) and greater care in the administering of antibiotics. Additionally, patients who contract a 'superbug' are often isolated from other patients to reduce the possibility of infecting others.

You need to be able to carry out experiments to investigate the effects of different types or strengths of antibiotics on bacterial growth, using aseptic techniques. The following question is based on this type of investigation.

Question

9 The following diagram shows how bacterial growth is affected by four different antibiotics, A, B, C and D.

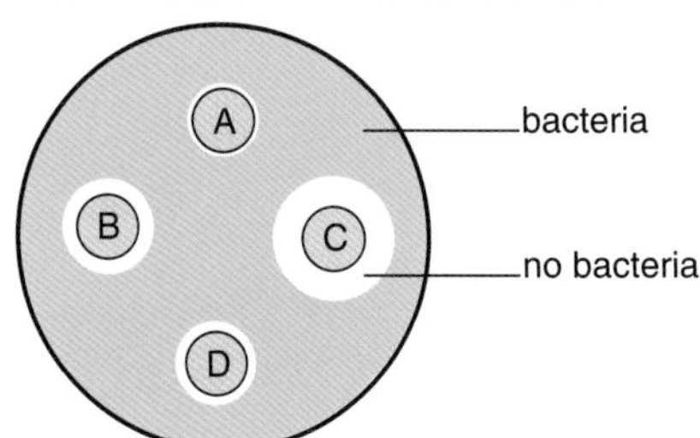

a) Which antibiotics were most and least effective in killing the bacteria?

The test was repeated a few months later using the same antibiotics on the same type of bacteria. The result is shown below.

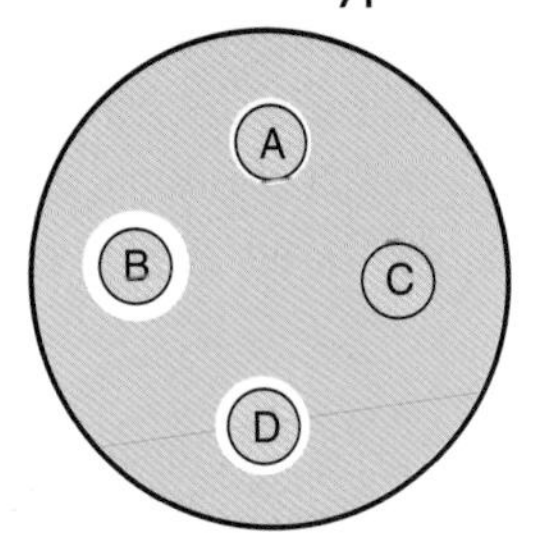

b) Describe and explain the difference between the results of the two tests.

c) Apart from using the same bacteria and the same antibiotics, state two other things that would need to have been done to make the results reliable.

Discovering medicines

Medicines have been around for a very long time (although not the same type that we get from the pharmacist today). Medicines are substances that help us recover from illnesses or reduce discomfort or pain. Rubbing dock on nettle stings to reduce pain is a very old remedy, as is the use of iodine to help heal cuts.

The discovery of medicines in earlier times was often by accident or chance. An example of this can be seen in one of the most famous medical discoveries of the last century – the discovery of **penicillin**.

Penicillin

Penicillin was the first antibiotic to be developed. In 1928 **Alexander Fleming** was growing bacteria on agar plates and noticed that one of his plates had become contaminated with mould (fungi). This is common when culturing bacteria unless a very effective aseptic technique is used. The growth of mould did not surprise Fleming, but he was surprised when he noticed that the bacteria he was culturing did not grow around the edges of the mould. He concluded that the mould produced a substance that prevented the growth of the bacteria. As the fungus causing the contamination was called *Penicillium*, the antibacterial substance was called **penicillin** and the first antibiotic was developed.

Figure 15.12a Florey

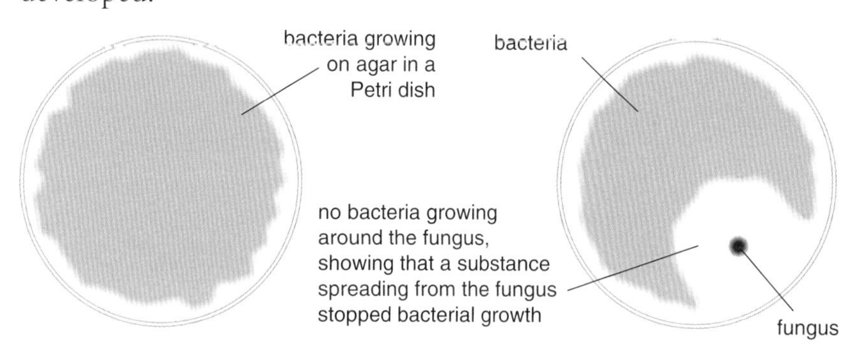

Figure 15.11 Fleming's discovery

Fleming carried out some work with his antibacterial substance on animals, but his progress was hindered because he was unable to produce a pure form of the substance. In the early 1940s two other scientists, Florey and Chain, were able to isolate a pure form of penicillin and its large-scale production began. Penicillin has been in use since then but is now only one of a large number of antibiotics in use.

Figure 15.12b Chain

Making penicillin commercially

Penicillin and other drugs are made in very carefully controlled conditions that maximise productivity. The microbes that make the penicillin are grown in large **biodigesters** or **fermenters** that create the perfect conditions for fungal growth. Downstreaming (extraction, purification and packaging) is required following the production of the penicillin (as with insulin as discussed in Chapter 12). Figure 15.13 shows a typical fermenter.

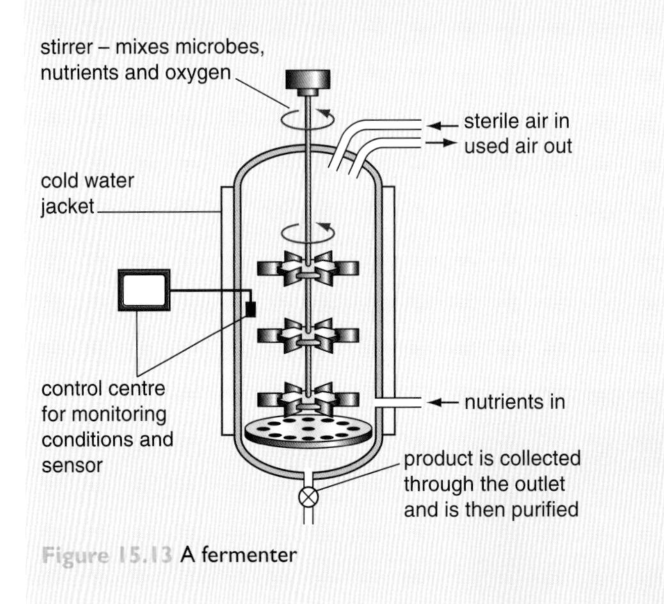

Figure 15.13 A fermenter

New medicines are continually being developed and the development of new drugs has made the pharmaceutical industry a major contributor to developed economies today.

Other drugs

There are other chemicals that some people take for reasons other than making them better when they are ill. These include **alcohol**, **nicotine** in cigarettes and so-called **recreational drugs** such as **cannabis** and **cocaine**.

Alcohol

Many people drink alcohol in moderation and are unlikely to suffer any serious harm. However, many people, including many teenagers, drink too much and cause harm to themselves and others.

Why do teenagers drink too much?

Some of the reasons include:

* peer pressure
* experimentation
* trying to escape from problems.

Harm caused by alcohol

Long-term excessive drinking can damage the **liver** as well as other parts of the body. Drinking heavily during pregnancy can cause serious damage to the foetus including brain damage. Drinking too much can harm those around you and society, as a result of:

* violence – many people become aggressive when drinking alcohol
* absence from college or work
* family breakup
* breakdown in relationships
* drink-driving.

Figure 15.14 A consequence of binge drinking

Binge drinking is a particular problem. This occurs when a large amount of alcohol is drunk over a short period of time, for example, on one night out (Figure 15.14).

The Government has tried to reduce the effect of binge drinking by extending the licensing hours. It is debatable whether this is having the desired effect!

What can we do to reduce the harm caused by alcohol?

There are many things that we can do.

* Drink less each time, by drinking low-alcohol drinks or just by drinking more slowly.
* Drink on fewer occasions, for example, only on special occasions or at weekends.
* Education – everyone should know how many units make up the recommended maximum weekly limit and about the problems alcohol can cause.
* Never drink and drive.
* Do not drink alcohol until you reach the legal age limit.

Smoking

Smoking can seriously damage health as summarised in the following table.

Substance in cigarette smoke	Harmful effect
Tar	Causes **bronchitis** (narrowing of the bronchi and bronchioles), **emphysema** (damage to alveoli that reduces the surface area for gas exchange) and **lung cancer** (caused by abnormal cell division in lung cells)
Nicotine	**Addictive** and affects **heart rate**
Carbon monoxide	Combines with red blood cells to reduce the **oxygen**-carrying capacity of the blood

The introduction of smoking bans in many countries has proved very effective. It both encourages smokers to stop and significantly reduces people being affected by passive smoking.

Question

10 a) Explain why people with bronchitis may find it difficult to be very active.

b) Why do many smokers find it so difficult to stop?

Figure 15.15 Some of the illegal drugs in use in the UK

Illegal drugs

Cannabis and cocaine are two of the most common illegal drugs used in Northern Ireland. Cannabis has been widely used, mainly by young people, for many years but the use of cocaine has risen rapidly in recent years (Figure 15.15).

Cannabis

* It is widely used throughout the UK, due to its availability and low cost.
* When taking cannabis, people may feel relaxed or 'chilled out'.
* However, it is possible that taking cannabis can lead to taking other, harder drugs. There is also evidence that cannabis can lead to mental health problems in some people.

Cocaine

* Cocaine can give users a 'high', as it is a stimulant.
* It is very addictive.
* As its effects are short-lived, users often increase their dose with time.
* An overdose can result in death.

It is important to remember that cannabis and cocaine are **illegal drugs**. Cocaine is a Class A drug, which the law says is the most dangerous category. The classification of cannabis has changed twice over recent years; it is now a Class B drug. There is still a lot of debate about the classification of drugs for legal purposes.

Exam questions

1 a) Jane carried out the following investigation.

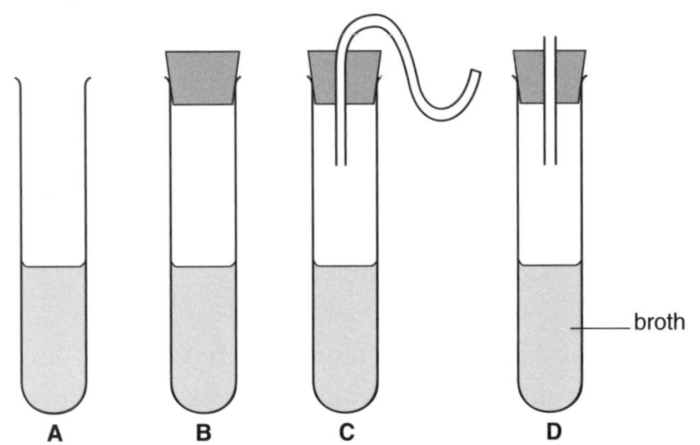

The apparatus and the broth were sterilised before the investigation and the experiment was left for a period of time.

i) Which tube(s) (**A**, **B**, **C** or **D**) will contain contaminated broth after a period of time? *(1 mark)*

ii) Name **one** type of organism that could have caused contamination in the broth. *(1 mark)*

iii) Name the scientist who used similar apparatus to investigate contamination. *(1 mark)*

iv) Copy the table below and tick the things that Jane must do to obtain valid results.

Action	Tick if essential to give valid results
Same volume of broth used in all tubes	
Same type of broth used in all tubes	
Tubes kept in the fridge during the investigation	
Tubes kept in bright light during the investigation	
Tubes kept at the same temperature during the investigation	
Tubes left for the same length of time during the investigation	

(2 marks)

b) Explain fully what is meant by the term aseptic technique. *(2 marks)*

2 The diagram shows the apparatus used by Pasteur in an experiment.

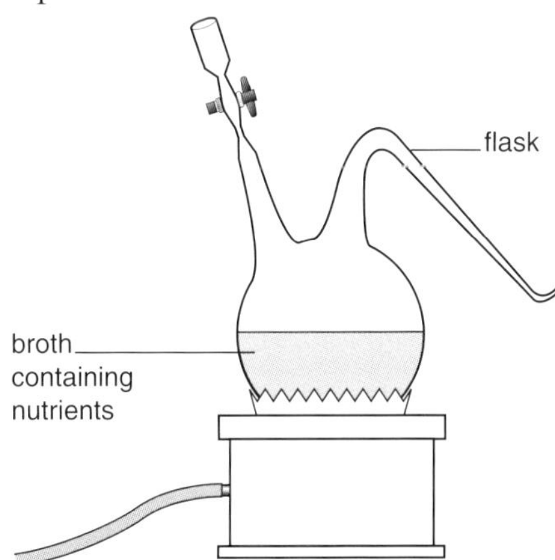

The broth in the flask remained clear for many weeks after boiling. It quickly turned cloudy when Pasteur tilted the flask until the broth reached the bend in the neck.
Explain these two observations and describe how Pasteur used this experiment to refute the theory of spontaneous generation. *(6 marks)*
In this question, you will be assessed on using your written communication skills including the use of specialist science terms.

3 a) i) Name **one** disease caused by bacteria. *(1 mark)*

The diagram shows how a blood cell called a phagocyte deals with an invading bacterium.

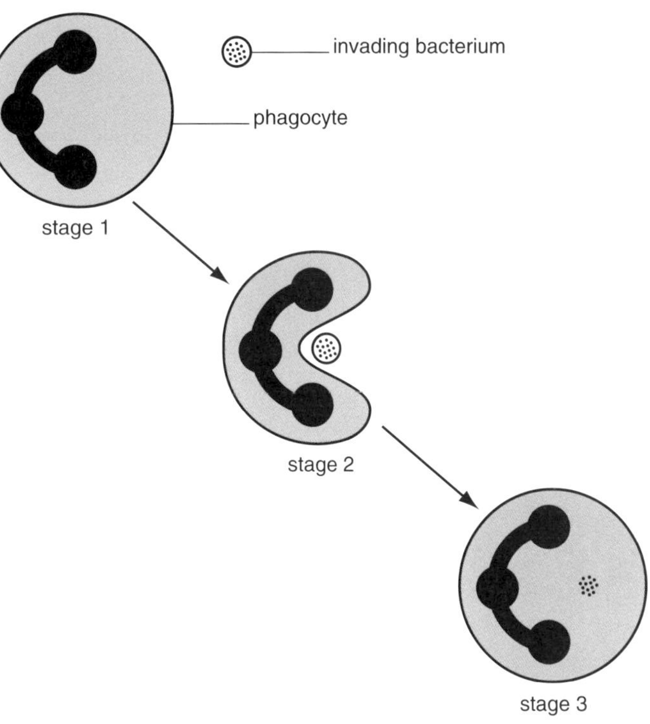

ii) What type of blood cell is a phagocyte? *(1 mark)*

iii) Describe what is happening at stages 2 and 3. *(2 marks)*

b) Another way of dealing with invading bacteria that cause disease is to produce antibodies. The diagram shows the sequence of events following the production of antibodies.

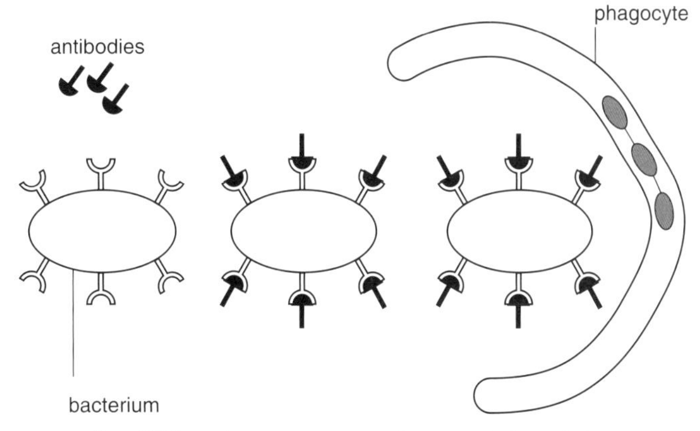

i) What is present on the outside of the bacterium that causes the antibodies to be produced? *(1 mark)*

ii) Use the diagram and your knowledge to explain why the antibodies shown in this diagram would not protect a person from invasion by a different bacterium. *(1 mark)*

iii) Suggest why someone is unlikely to get the same disease again, once they have produced antibodies against a particular type of bacteria. *(2 marks)*

4 a) Describe how the mucous membranes in the nasal cavity (nose) protect against disease. *(2 marks)*

b) The diagram below shows how the antibody level in the body changed after a vaccination. It also shows that immunity was not achieved.

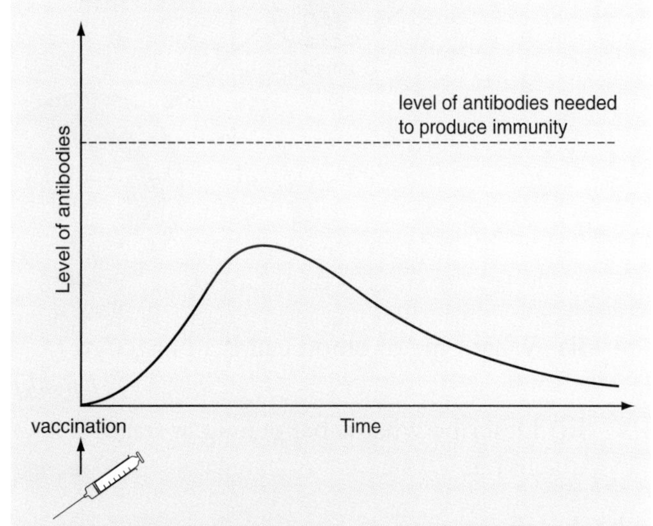

i) Suggest **one** thing that should have been done to ensure that enough antibodies were produced to achieve immunity. *(1 mark)*

ii) Describe and explain how a vaccination works. *(3 marks)*

c) Explain why someone travelling to Africa for the first time may be given travel vaccinations before they go. *(2 marks)*

5 a) Copy and complete the flowchart showing different forms of immunity using words from the list below.

antibody acquired active
foreign natural antigen

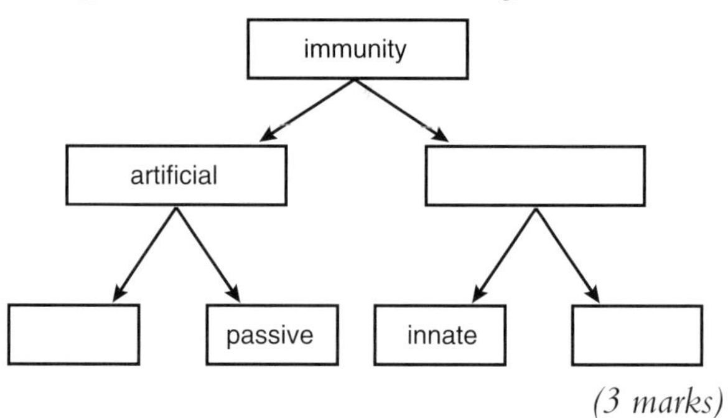

(3 marks)

b) A doctor was asked at short notice to travel abroad to help rescue survivors of an earthquake.

To protect him against a disease, he was given an injection that provided him with artificial, passive immunity.

The table shows the level of antibodies in his blood over the next 15 weeks.

Time since injection/ weeks	Level of antibodies in blood/arbitrary units
0	0
1	20
3	18
5	16
7	13
9	9
11	5
13	2
15	0

Use evidence from the table to help explain artificial, passive immunity. *(4 marks)*

c) Explain how the antibodies in the injection and phagocytes in the doctor's blood would protect him if he was infected by the disease microorganism. *(4 marks)*

The minimum level of antibodies in the blood that protect against the disease is 14.5 arbitrary units.

d) i) How long would it be safe for the doctor to stay abroad? *(1 mark)*

ii) Explain why the doctor should return home before this time. *(1 mark)*

6 The following diagram shows what Alexander Fleming observed when an agar plate containing bacteria was contaminated by a fungus.

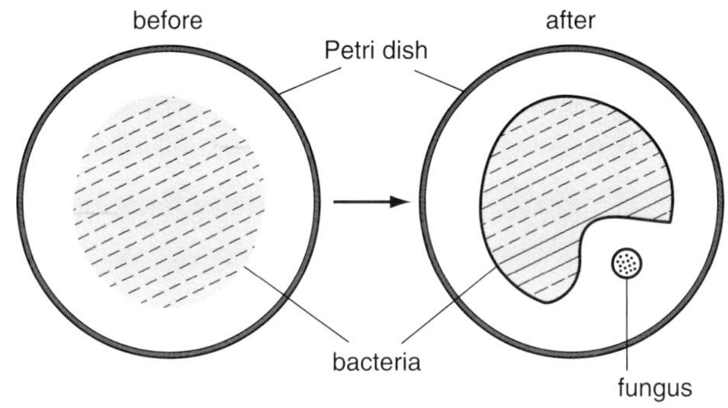

a) Describe fully what Fleming concluded from this observation. *(2 marks)*

b) In a laboratory experiment Kathryn transferred bacteria from tube A to a Petri dish. She attempted to use aseptic techniques but her method was incorrect and produced contaminated results. Given below is an account of her method.

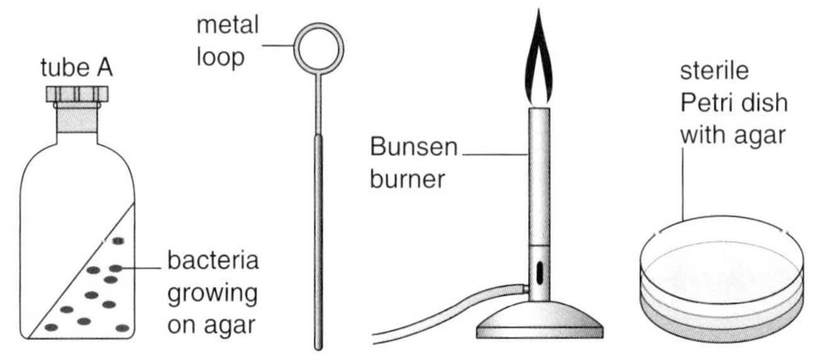

Kathryn removed the lid from the Petri dish and left it to one side. She then held the metal loop in the Bunsen flame for 30 seconds. While it was still hot she dipped it into tube A and transferred the bacteria, spreading it over the agar in the Petri dish. She replaced and sealed the lid on the Petri dish and incubated it at 20 °C for 2 weeks. When Kathryn opened the Petri dish she found that none of the bacteria from tube A had grown, but other types of microbes were growing on the agar.

i) Why is it important that the Petri dish is sterilised before the start of the experiment? *(1 mark)*

ii) Suggest **one** reason why the bacteria from tube A had not grown. *(1 mark)*

iii) Suggest **one** reason why the Petri dish had become contaminated. *(1 mark)*

7 The diagram shows a simple biodigester used to produce penicillin.

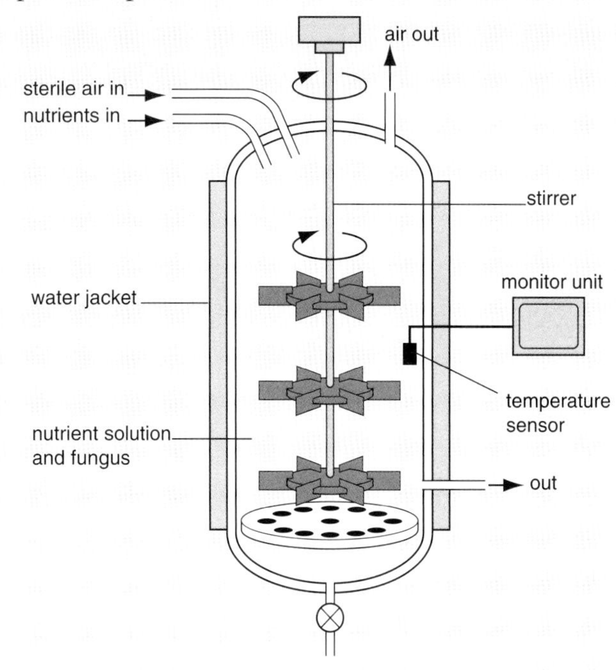

a) What is the function of the stirrer? *(1 mark)*

b) Explain how the biodigester prevents the temperature rising too high. *(2 marks)*

c) Explain why the temperature of the biodigester should not be allowed to rise too high. *(1 mark)*

d) Explain why the air entering the biodigester must be sterile. *(1 mark)*

e) Describe what must happen to the contents of the biodigester before the penicillin can be used. *(3 marks)*

8 a) Smoking tobacco, drinking alcohol and the misuse of drugs causes harm to people and society. The following table compares the effects of these substances on individuals.

Key Most effect = 3
Least effect = 1

Substance	Can cause mental health problems	Can cause road traffic accidents	Can cause cancer
Tobacco	1	1	3
Alcohol	3	3	2
Cannabis	2	2	1

i) Tobacco and alcohol can legally be used by adults, but cannabis remains an illegal drug. Use the information in the table to explain how campaigners could suggest that taking cannabis should be made legal. *(2 marks)*

ii) Suggest why it is illegal for children to purchase cigarettes and alcohol, but legal for adults. *(1 mark)*

iii) Name the substance in tobacco that:
1. is addictive *(1 mark)*
2. causes cancer *(1 mark)*

b) The following table shows the pattern of drinking of three teenagers over a 7 day period. The values show the number of alcoholic units taken.

	Mon	Tue	Wed	Thurs	Fri	Sat	Sun	Total
John	1	2	3	2	2	2	0	
Tony	0	0	2	0	0	8	0	
Paul	0	0	0	0	2	2	0	

i) Copy and complete the table to show the total number of units each teenager had taken during the week. *(1 mark)*

ii) Name the teenager who was binge drinking. *(1 mark)*

iii) Suggest **two** reasons why the Government recommends that people do not binge drink. *(2 marks)*

Controlled Assessment

As part of your GCSE course you will have to complete a controlled assessment.

These controlled assessment activities will be based on, or similar to, experiments you should have done as part of your course. For example, the practical activity outlined in the CCEA Specimen Paper for both Double Award Science and GCSE Biology is testing the vitamin C content of a range of juices.

Each controlled assessment has three parts:

* Part A – Planning and Risk Assessment
* Part B – Data Collection
* Part C – Processing, Analysis and Evaluation

Part A: Planning and risk assessment

Planning

In this section you should plan how you are going to carry out the task and state a **hypothesis** (what you think your results will show). Remember, your hypothesis is a prediction – what you predict may not necessarily happen. This is not a problem as long as your prediction is a reasonable one based on the background information you bring to your planning.

To gain full marks in this section you need to be able to develop a hypothesis but also explain why you have produced your hypothesis. You should use your scientific knowledge and terminology relevant to the topic.

The plan you devise will need to be **complex** and **valid**, meaning it will test what you are asked to test and will allow you to accept or reject your hypothesis. It also needs to produce **reliable** results. Reliability is affected by both the **range** of results you plan to take and also the **number** of results (including repeats where appropriate).

In addition, you must explain how you would deal with any **anomalous results** (should they arise). The form, style, spelling, grammar and punctuation (quality of written communication) should also be of a high standard.

Take for example an enzyme experiment investigating the effect of amylase concentration on the rate of starch breakdown. Testing the effect of just two concentrations, for example 10% and 50% amylase, would not be a wide enough range to give reliable results. A range of five values, e.g. 20%, 40%, 60%, 80% and 100%, would be more appropriate. In addition, repeating the experiment to obtain a second result for each concentration would help its reliability.

You should state clearly the **variable** you are going to manipulate (the **independent variable**). In the above example this is the concentration of amylase. You also need to state clearly how you will obtain your results – measure the **dependent variable**. In the enzyme example this is the time taken for the amylase to break down the starch. You will obviously need to describe how you set the experiment up to obtain the time, for example the use of spotting tiles containing iodine that the starch/amylase solutions are added to at time intervals of 30 seconds.

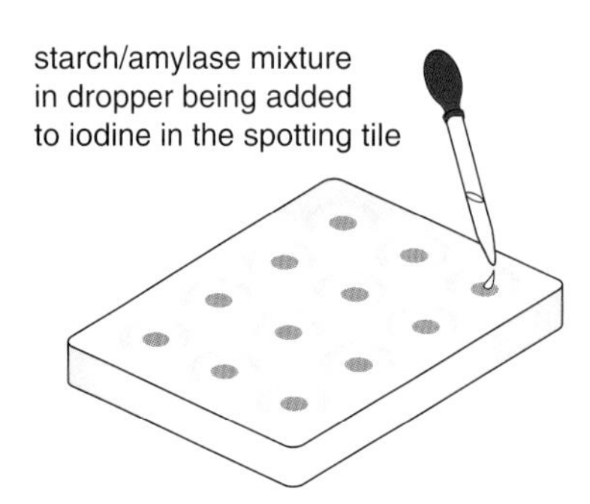

Figure 16.1

It is important that *other variables are controlled*. In the amylase enzyme experiment it is important to keep a number of variables constant, such as volume of enzyme, volume and concentration of the starch solution, and temperature. If any of these were not controlled then you could not be sure you had obtained valid results. This is because you could not be certain that it was the change in enzyme concentration that produced the change in rate of starch breakdown. As well as knowing which variables to control you need to explain how you will keep them controlled, for example by using a water bath to keep temperature constant.

You will need to describe the **apparatus** and **equipment** that you will require. Remember that you should select apparatus that will give you the degree of precision and accuracy you need. In the amylase experiment you would probably use a syringe or a graduated cylinder to measure volumes.

You also need to have:

* a detailed method that describes everything you are going to do and the order in which you are going to do it
* a blank table for your results. This should have headings (and units if appropriate).

Risk Assessment

Many practical activities carry an element of risk. However, you are unlikely to be set a task that is dangerous for students to carry out. You need to identify the safety hazards *specific to the investigation you are planning* and the apparatus you will be using. These will include obvious generic safety precautions that apply to many experiments but possibly also very specific precautions that apply to only one type of investigation.

In an enzyme experiment your risk assessment could include making sure all solutions are kept away from electrical sockets on the bench, wearing safety glasses to avoid splashes, etc. You need to both identify the hazards and describe how you will minimise risk.

The planning and risk assessment is marked by your teacher. You must complete this section in school/college as you will not be allowed to take it home. This section is worth 18 marks out of a total CAT allocation of 45 marks.

Part B: Data Collection

If your plan is appropriate and detailed enough, you will be able to use it (possibly with minor suggested changes by your teacher) to obtain the results you need.

However, if your plan has significant problems – for example, if it might be dangerous or if you may not get the results you need – you could be given a new plan by your teacher. The advantage of this is that you can still gain full marks in Part C (where most of the marks are available) even if your Planning and Risk Assessment section has some limitations.

There are no marks available for the Data Collection section. However, this section is important as you will be asked to analyse and interpret the data you have obtained in Part C.

Part C: Processing, Analysis and Evaluation

This part is completed in a separate booklet (Candidate Response Booklet B) from the earlier sections and it is very like an ordinary examination. Your school/college will treat it like an examination – you must complete this section entirely without help and the booklet must be handed in at the end of the session.

In this section there are a number of things you are very likely going to be asked to do. These include:

* drawing a **graph** to display your results – to gain full marks you will probably be asked to scale and label your axes
* **describing** and **explaining** what your results show, including whether or not your results support your hypothesis
* identifying any **trends** that your results show
* identifying any **anomalous** results
* discussing how you ensured that your investigation provided **valid** and **reliable** results
* carrying out some **calculations**
* suggesting some **improvements** to your method.

You will also be provided with other data (this is called 'secondary data' as it is not data you have obtained) that may relate to the type of investigation you have carried out. This data may be in the form of tables or graphs. Typically there will be three graphs or tables.

In our enzyme example you could be given the following three graphs:

Graph 1 – a graph of the time taken to break down starch against volume of enzyme

Graph 2 – a graph of the time taken to break down starch against temperature

Graph 3 – a graph of the time taken to break down starch against pH.

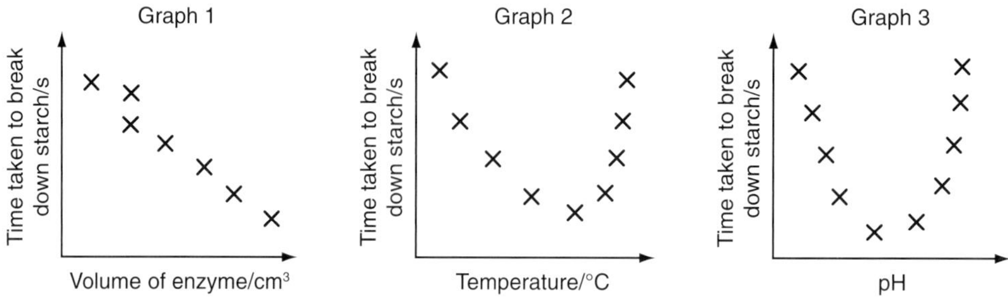
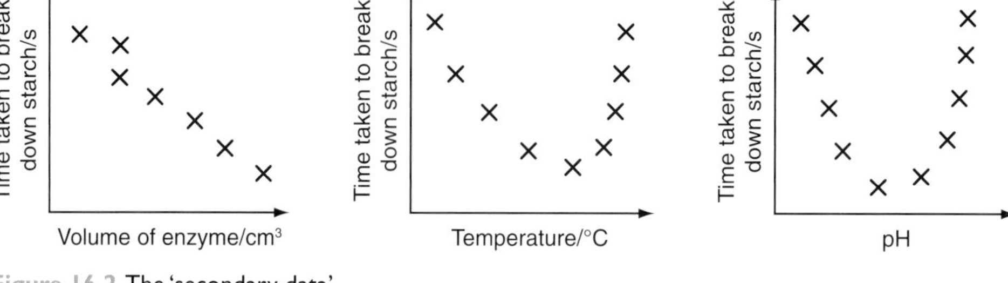

Figure 16.2 The 'secondary data'

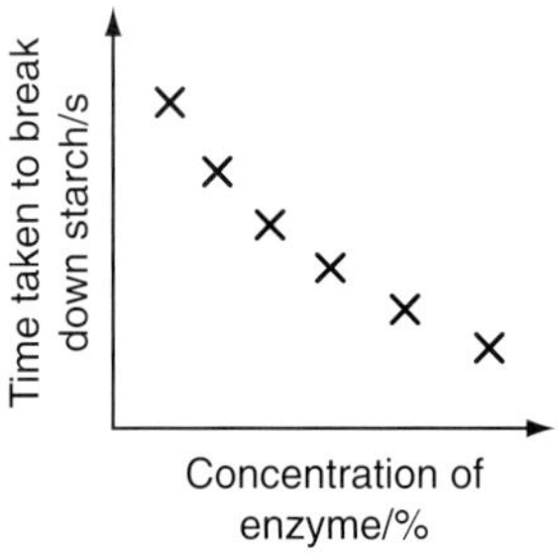

Figure 16.3 Your results

You will be expected to analyse the secondary data and compare it with your results. The type of questions you could be asked include:

* Which of the graphs provided most closely resembles your own? *(graph 1)*
* Are there any anomalous results in the graphs provided? *(see graph 1)*
* Account for any differences between the graph that most closely resembles your own and your own results. (*if in your experiment the starch was broken down much more quickly than in Graph 1 this could be due to a number of factors, e.g. you have used more enzymes, your experiment was closer to the optimum temperature, etc.*)
* You could also be asked to write an account summarising the conditions in which amylase will break down starch fastest. In this you would be expected to relate to both the secondary data and your own results. (*Your answer should make reference to how volume of enzyme, temperature, pH* ***and*** *enzyme concentration affect the rate of enzyme activity.*)

The investigation(s) you will be asked to do as your CAT varies from year to year. However, you should be able to work out the type of investigation you are likely to be asked to do. It will be one where you have an opportunity to make a risk assessment, plan a complex investigation, obtain a range of results, and consider methods of ensuring validity and increasing reliability.

Index

abiotic factors 53, 54–5, 68
acid rain 66–7
active immunity 139, 140
active transport 37, 65
adaptations 121
aerobic respiration 36–7, 65
air pollution 68
alcohol 146–7
alimentary canal 26–8
alleles 95–6, 97, 100
amino acids 23, 82–3
amniocentesis 114
amylase 26, 29
anaemia 130
anaerobic respiration 37
animal cells
  growth 5
  osmosis in 75
  structure 1
animals
  classification 56
  distribution 52–3, 55
antibiotics
  bacterial resistance 122, 143
  diseases and 142–4
  penicillin 145–6
antibodies 137, 138–9, 141
antigens 137, 138, 141
anus 26, 28
apices 5
aqueous humour 40
arteries 127
arthritis 21
aseptic techniques 141–2
asexual reproduction 88
auxin 46–7

bacteria
  antibiotic resistance 122, 143
  as cause of disease 135, 136
  cells 3
  classification 56
  decomposition by 61, 65
balanced diet 20, 22–4
basal metabolic rate (BMR) 20
bases 81–3
behaviour 39
belt transects 52, 54–5
benign tumours 90
bile 26, 27
biodiversity 53
biotic factors 53, 68
blood 126–8
  clotting 126, 137
  donation 130–1
  glucose in 44–6
blood vessels 127, 131
bloodworms 68
body mass index (BMI) 19–20
brain 21, 39, 129
breathing 33–5, 36
bubble potometers 76–7
buccal cavity 26

cancer
  causes 89, 113, 147
  screening 90
  treatment 91
  types of tumour 90
cannabis 148
capillaries 127, 131
carbohydrates 22, 27
carbon cycle 62–4
carbon dioxide
  global warming and 62–3
  in photosynthesis 7, 9, 11, 13, 14–15
  reducing emissions 68
cell division
  meiosis 91–3, 99, 106
  mitosis 87–8, 93, 106
cell membranes 1, 4, 74
cells
  animal *see* animal cells
  bacterial 3
  blood 75, 126, 137–8
  differentiation 5
  egg cells 99, 106
  movement of substances in 4, 74–6
  observing 2–3
  organisation 4–5
  plant *see* plant cells
  specialised 5, 65
  sperm cells 99, 106
  stem cells 5
  structure 1
cell walls 1, 74
central nervous system 39
Chargaff, Erwin 83
chemotherapy 91
chlorophyll 7, 9
chloroplasts 1
chromosomes 1, 81, 82, 95, 96
  cell division and 87–8, 91–3
  sex chromosomes 99–101
ciliary muscle 41
circulatory system 126–31
classification 56–7
cloning 87, 88
clotting 126, 137
cocaine 148
colon 26, 28
colour-blindness 100–1
combustion 62, 63, 66
community 56
compensation point 14
conjunctiva 40
continuous variation 120, 121
contraception 108
controlled assessment 153–6
cornea 40
coronary arteries 128, 129
Crick, Francis 84
cuticle 11
cystic fibrosis 98–9, 114
cytoplasm 1

Darwin, Charles 122
decomposition 61, 62
deforestation 63
denitrification 65
diabetes 21, 45–6, 115–17
differentiation 5
diffusion 4, 74
digestion 26
digestive system 4, 26–8
diploid number 92, 106
discontinuous variation 121
diseases
  caused by microbes 136
  defence against 137–41
  inherited conditions 21, 98–9, 114–15
  vaccinations 139–41
  see also cancer; diabetes; heart disease
distribution
  abiotic factors 53, 54–5, 68
  sampling 51–3, 54–5
DNA
  function 82–3
  genetic engineering and 115–17
  structure 81–2, 83–4
dominant alleles 95, 97
downstreaming 117, 146
Down syndrome 113–14
drugs
  alcohol 146–7
  illegal 148
  medicines 142–6
duodenum 26, 27

ecology 51
ecosystems 56, 57
effectors 39
egg cells (ova) 99, 106
embryos 5, 106, 110
endangered species 123
energy
  in ecosystems 57
  energy flow 57, 61
  in food chains 60–1
  from food 18, 22, 60–1
  from respiration 37
  renewable 68
  requirement for 18–19
  sources 7

environment 56
  change monitoring 68
  Government role in conservation 68–9
enzymes
  action of 28–30, 116
  commercial 30
  in decomposition 61
European Nitrates Directive 69
eutrophication 66
evolution 122
exercise 36, 130
extinction 123
eyes 40–2

fats 20, 22, 23, 27
female reproductive system 106–7
fertilisation 92, 106
fertilisers 66, 69
fertility problems 109–10
fibre 24
Fleming, Alexander 145
focusing 40–1
foetus 106–7
food
  energy content 18, 22
  health and 20–1, 22–4
  tests for 21–2
food chains 57–8
  energy in 60–1
food webs 58–9
fossil fuels 62, 63, 66
Franklin, Rosalind 84
fruit
  in diet 20
  production 47–8
fungi
  as cause of disease 135, 136
  classification 56
  decomposition by 61, 65

gall bladder 26, 27
gametes 91–2, 106, 113
gas exchange
  in leaves 11–12, 36
  in respiratory system 33, 35–6
genes 81, 95
genetic engineering 115–17
genetic profiles 115
genetics 93–8
genetic screening 114–15
genotypes 95
global warming 62–4
glucagon 45
glucose
  in the blood 44–6
  in food 22
  from photosynthesis 7, 12
Government
  alcohol abuse and 147
  environmental issues and 68–9
  stem cell research and 5

growth
  measuring 87
  patterns 5

habitat 56
haemoglobin 126
haemophilia 100–1, 113
haploid number 92, 106
Harris Benedict equation 20
health
  circulatory system and 129–30
  food and 20–1, 22–4
  see also diseases
heart 128
heart attacks 129
heart disease 21, 46, 129
height 19–20, 87, 120
heterozygosity 95, 97, 100
high blood pressure 21
homozygosity 95, 97, 100
hormones
  human 43–6
  plant 46–8
human papilloma virus (HPV) 89
humidity, transpiration and 77
humus 61
hypothesis 153

ileum 26, 27–8
illegal drugs 148
immunity 138–9, 140
indicator species 68
inheritance 21, 93, 96–7
insulin 44–5, 115–17
in vitro fertilisation 110
iris 40, 41–2

Jenner, Edward 140

karyotypes 99, 113–14
Kyoto Protocol 69

large intestine 26, 28
law of segregation 95
leaves
  diffusion 4
  gas exchange 11–12
  light absorption 10–11
  photosynthesis 7–9
  structure 11
  transpiration and 77
  variegated 9
  see also plants
legumes 65
lens 40–1
lichens 68
light
  distribution and 53
  in the eye 41–2
  photosynthesis and 7, 9, 10–11, 12, 13
  plant response to 46–7
limiting factors 12–13

lipase 27
liver 26, 27
lock and key model 29
lungs 33–5
lymph 131
lymphocytes 126, 137
lysis 75

magnification 3
malignant tumours 90
mass 19–20, 87
medicines
  antibiotics 142–4, 145–6
  discovery of 145–6
meiosis 91–3, 99, 106
memory lymphocytes 138
Mendel, Gregor 93–6
menstrual cycle 109
menstruation 109
microorganisms
  contamination by 135–6
  disease and 136–41
  in the laboratory 141–2
microscopes 2
minerals 23
mitosis 87–8, 93, 106
monohybrid crosses 94, 96–7
mouth 26
MRSA 143
mutations 113

natural selection 121–2
negative feedback 45
nerve impulses 39
nervous system 4, 39–43
nets 53
neurones 39, 42–3
nicotine 146, 147
nitrifying bacteria 65
nitrogen cycle 64–5
nitrogen fixation 65
nuclear membrane 1
nucleotides 82
nucleus 1, 81
nutrient cycles 61–6

obesity 19, 20, 21
oestrogen 43, 109
optic nerve 40
organisms 4
organs 4–5
organ systems 4–5
osmosis 74–6
ovulation 109
oxygen
  from photosynthesis 7, 11
  plant respiration 14–15, 65

palisade mesophyll 11
pancreas 26, 27, 44, 45
passive immunity 139, 140
pasteurisation 136

Pasteur, Louis 135–6
pedigree diagrams 98–9
peer review 5, 48, 84
penicillin 145–6
percentage cover 51
pH
  digestion and 27
  distribution and 53
  enzyme action and 29–30
phagocytes 126, 137
phagocytosis 126, 137–8
phenotypes 95
photosynthesis 7, 62
  experiments 7–9
  maximising 13–14
  rate of 7, 10, 12–13
  respiration and 14–15
phototropism 46–7
pitfall traps 52
placenta 106, 107
plant cells
  growth 5
  organisation 4
  osmosis 75–6
  root hair cells 65
  structure 1
  transpiration 76–8
plants
  asexual reproduction 88
  classification 56
  distribution 51–2, 54–5
  phototropism 46–7
  plant hormones 46–8
  respiratory surfaces 36
  sensitivity 46–7
  see also photosynthesis
plasma 126, 131
plasmids 3
plasmolysis 75
platelets 126
pleural fluid 35
pleural membranes 35
pollution 66, 68, 69
pooters 53
population 55, 56
potometers 76–8
pregnancy 19, 106–7, 109
primary consumers 57, 58
producers 57, 58
protease 27
protein 22, 23, 27, 64
pulse rate 130
Punnett squares 96–7
pupil 40, 41–2
pupillary reflex 42
putrefying bacteria 65
pyramids of biomass 60
pyramids of numbers 59

quadrats 51–2

radiotherapy 91
receptors 39
recessive alleles 95, 97
recreational drugs 146, 148
rectum 26, 28
red blood cells 75, 126
reduction division *see* meiosis
reflex actions 42–3
reliability 51–2, 55, 77, 153
renewable energy sources 68
reproduction
  asexual 88
  sexual 106–7
reproductive system 4, 106–7
respiration 62
  aerobic 36–7, 65
  anaerobic 37
  glucose and 12
  photosynthesis and 14–15
  in plants 14–15, 65
respiratory system 33
  breathing 33–5, 36
  surfaces 35–6
response 39
restriction enzymes 116
retina 40
risk assessment 154–5
root hair cells 65

sampling 51–3, 54–5
screening programmes 90
secondary consumers 57, 58
secondary sexual characteristics 109
sex chromosomes 99–101
sex-linked conditions 100–1
sexual reproduction 106–7
slides 2–3
small intestine 26, 27–8
smoking 89, 147
specialised cells 5, 65
sperm cells 99, 106
spinal chord 39
spontaneous generation 135
starch
  in food 22, 26
  in photosynthesis 7–8
stem cells 5
'sticky ends' 116–17
stimuli 39
stomach 26, 27
strokes 21, 46, 129
sugar 20, 22, 26
  see also glucose
sulphur dioxide 66–7
superbugs 143–4
survival of the fittest 121, 122
suspensory ligaments 41
synapses 43

temperature
  distribution and 53
  enzyme action and 29
  photosynthesis and 12
  transpiration and 77
tertiary consumers 57, 58
test cross 97–8
testosterone 43, 109
tissue culture 88
tissue fluid 131
tissues 4–5
transpiration 76–8
trophic levels
  energy transfer between 60–1
  in food chains 57, 58
turgor 75

umbilical cord 106, 107
uterus 106, 107
UV radiation, cancer and 89, 113

vaccinations 139–41
vacuole 1
validity 55, 77
variables 153–4
variation 120–1, 122
variegated leaves 9
vegetables 20
veins 127
villi 27–8, 107
viruses
  as cause of disease 89, 135, 136
  classification 57
vitamins 22, 23
vitreous humour 40
voluntary actions 42

water
  in the body 24, 28
  distribution and 53
  osmosis 74–6
  in photosynthesis 7
  pollution 66, 68
  transpiration 76–8
Watson, James 84
weedkillers 47
weight *see* mass
white blood cells 126, 137–8
Wilkins, Maurice 84
wind
  distribution and 53
  transpiration and 77

X-ray diffraction 84

yeast 37

zygotes 87, 106